HEAT TRANSFER SOLVER

HEAT TRANSFER SOLVER

M.D. Mikhailov
North Carolina State University
Raleigh, North Carolina

M.N. Özisik
Applied Mathematics Center
Sofia, Bulgaria

Prentice Hall
Englewood Cliffs, New Jersey 07632

Editorial/production supervision: ***Carolyn Del Corso***
Cover design: ***Carolyn Del Corso/Benjamin D. Smith***
Manufacturing buyers: ***Linda Behrens/Dave Dickey***
Supplement acquisitions editor: ***Alice Dworkin***
Acquisitions editor: ***Douglas Humphrey***

A Division of Simon & Schuster
Englewood Cliffs, New Jersey 07632

Printed in the United States of America

10 9 8 7 6 5 4 3 2 1

ISBN 0-13-388802-9

Prentice-Hall International (UK) Limited, *London*
Prentice-Hall of Australia Pty. Limited, *Sydney*
Prentice-Hall Canada Inc., *Toronto*
Prentice-Hall Hispanoamericana, S.A., *Mexico*
Prentice-Hall of India Private Limited, *New Delhi*
Prentice-Hall of Japan, Inc., *Tokyo*
Simon & Schuster Asia Pte. Ltd., *Singapore*
Editora Prentice-Hall do Brasil, Ltda., *Rio de Janeiro*

CONTENTS

PREFACE

During the past decade the use of personal computers has increased significantly and more recently almost all engineers and students have access to personal computers. This software and its accompanying manual are intended to serve as a useful tool for practicing engineers and engineering students to perform heat transfer calculations on fourteen different topics covered here. It can be used by students to supplement the standard undergraduate level heat transfer book.

Anyone who is familiar with the physical aspects of heat transfer problems and how to choose the input data, can use this software to solve practical heat transfer problems without a need for a formal background in the mathematical aspects of heat transfer.

Unique features of this software include, among others, the following:

* SCREEN EDITOR which allows the entire input data to remain on the screen all the time, thus avoiding scroll. Any data on the screen can be readily changed by moving the cursor over the data and entering the new value.
* MIXED UNITS allows the use of British and the SI system of units. At any instant the units of any input or output data on the screen can be readily changed by moving the cursor over the unit and pressing the spacebar which automatically alters the units.
* PARAMETER CHECKING displays a warning message on the screen whenever the permissible ranges of the empirical correlation is violated as a result of improper input data.
* GRAPHICAL SUPPORT displays the coordinate axes, automatically sets the units and some ranges for the variables. The user can edit the units as well as the end values of the variable intervals. This procedure is equivalent to ZOOMING in such a way that any portion of the graph can be magnified in the most convenient coordinate presentation.
* TABULATION of the exact data for up to 150 points on the curve. In the graphic mode, the values of the variables corresponding to the cursor location on the graph are displayed at the upper right corner of the screen. By moving the cursor along the

graph and pressing the ENTER key at each cursor point, the results corresponding to such cursor locations can be stored in the tabulated form.

* PRINT SUPPORT permits to retrieve a hard copy of the stored data in the tabulated form. A hard copy of the graph is also obtainable by using PrtSc key if GRAPHICS.COM is already entered.
* HELP is available at anytime by simply pressing the ? key.
* COLOR adjustments allows the user to alter background and/or foreground colors.
* The program is designed for the IBM-PC or compatible computer.

The software can be helpful for systematic investigation of the effects of various system parameters and material properties on heat transfer.

This work has been performed in the Mechanical and Aerospace Engineering Department of the North Carolina State University. We wish to acknowledge the assistance of Drs. M. Aladgem and M. Lavchiski in the preparation of the software.

M. D. Mikhailov M. N. Özisik

Raleigh, N. C.

OVERVIEW

The main features of each of the fourteen programs in this software are summarized.

1. STEADY STATE CONDUCTION calculates heat conduction without energy generation for a plate, hollow cylinder and hollow sphere subject to prescribed temperature boundary conditions at both surfaces. The program allows any of the physical quantities, such as thickness, length, inner and outer diameter, thermal conductivity to be an OUTPUT while all the others remain as INPUTS. The program can also handle the problems with temperature dependent thermal conductivity by using the Effective Thermal Conductivity concept as described in the text. The cases involving convection boundary conditions are treated in Chapter 2 as a special case of the composite medium problems.

2. COMPOSITE MEDIUM calculates one-dimensional, steady state heat flow rate and interface temperatures in composite plates, coaxial hollow-cylinders and concentric hollow-spheres of up to 9 layers in perfect thermal contact and subjected to prescribed temperature, prescribed heat flux or convection at the outer boundary surfaces. The contact resistance at any layer can readily be included by treating contact resistance as a fictitious layer. The graphic support plots the temperature distribution at all layers or at any specified interval over the layers. The special case of one layer medium is equivalent to the problem of a single layer solid.

3. LUMPED ANALYSIS calculates the time variation of temperature in solids in which the spatial variation of temperature is considered negligible. In the present problem, the solid is initially at a uniform temperature. For times $t>0$, it is exposed to convective heat transfer with an ambient at a constant temperature. The output can be either the time required for the solid to reach a specified temperature or the temperature of the solid at a specified time. The Graphical support plots the temperature as a function of time over a specified time interval.

4. TRANSIENT CONDUCTION calculates one-dimensional temperature as a function of time and position in a semiinfinite medium, a plate of finite thickness, solid cylinder and solid sphere initially at a uniform temperature. For times $t>0$ the boundary surface is subjected to a prescribed temperature, prescribed heat flux or convection. In the case of plate, both surfaces are considered subjected to the same boundary conditions. The graphical support presents the temperature at any specified location as a function of time over any given time interval.

5. FINS calculates the fin efficiency, heat loss per fin for longitudinal, radial and spine type fins of various cross-sections. Seven different types of fins are considered. The graphical support gives the temperature distribution along the fin.

6. FORCED CONVECTION calculates the average heat transfer coefficient for both laminar or turbulent forced flow inside a circular tube or a parallel-plate channel. In the case of laminar flow, the average heat transfer coefficient for thermally developing, hydrodynamically developed flow under constant wall temperature conditions are determined from the exact analytic solution of the classical Graetz problem. For the case of turbulent flow, thermally and hydrodynamically developed flow is considered. The standard recommended correlations are used to determine the heat transfer coefficient for flow over bodies and tube bundles. The graphical support allows the heat transfer coefficient to be plotted as a function of the flow velocity over a specified range of velocity.

7. FREE CONVECTION calculates the free convection heat transfer coefficient for vertical plate, vertical or horizontal cylinder and a single sphere. Graphical support allows the heat transfer coefficient to be plotted against the characteristic length.

8. FILMWISE CONDENSATION calculates the filmwise condensation heat transfer coefficient for condensation on a vertical plate, vertical tube, horizontal single tube and horizontal tube banks for both laminar and turbulent flow regimes. The graphical support is available to plot the heat transfer coefficient against the characteristic length.

9. NUCLEATE BOILING calculates the heat flux and the peak heat flux in the nucleate boiling regime in pool boiling for geometries such as flat plate, large finite body, horizontal cylinder and sphere. The graphical support presents the surface heat flux plotted against the temperature difference .

10. FILM BOILING computes the heat transfer coefficient for the film boiling regime including the effects of radiation. Graphical support is provided to plot the heat transfer coefficient against the characteristic length

11. BLACKBODY RADIATION FUNCTION plots the blackbody spectral radiation flux at a specified temperature as a function of wavelength over any given wavelength interval and displays the results in the graphical form. The values of the blackbody spectral flux are also displayed on the screen for selected wavelengths.

12. VIEW FACTORS calculate diffuse view factor for some simple geometries, including two equal parallel opposed rectangles, two coaxial parallel discs, and perpendicular rectangles with a common edge. Graphical support is given to display the view factor as a function of typical characteristics dimensions.

13. RADIATION SHIELDS calculates the heat transfer rate between two parallel plates, long coaxial cylinders and concentric spheres with up to 8 shields placed between them as well as displays the shield temperatures. The graphical support allows the plotting of the heat transfer rate through the system against the temperature of either the first plate or the second plate.

14. HEAT EXCHANGER EFFECTIVENESS calculates either the effectiveness or the heat transfer surface area for certain type of heat exchangers, including the double-pipe, shell-and-tube and cross-flow heat exchangers. The graphical support allows the plot of effectiveness against the heat transfer surface area of the heat exchanger.

TUTORIAL

The hardware needed for the operation of the system and the general procedure for starting, data entry, using graphical support and producing hard copies are now described. However, there are some variations in the application of this general procedure. Therefore, for each of the fourteen chapters, the first example gives a detailed description of data entry procedure and the use of graphical support for solving the problems belonging to that chapter. Therefore, it is advisable that the reader should go over the first example in each chapter before proceeding to the others.

HARDWARE REQUIREMENT

1. An IBM-PC, IBM-XT or IBM compatible computer with a 256 memory.
2. One disk drive at least.
3. An 80-column monitor: Color or monochromatic.
4. Printer (optional).

GETTING STARTED

1. Boot the computer.
2. Use the prompt A> if the Solver Diskette is inserted in drive A; or the prompt C> if the program is stored in the memory of the hard disc.
3. Activate the GRAPHICS mode by typing graphics if graphical support will be needed.
4. Type HTS (i.e., Heat Transfer Solver) and then press the ENTER (or RETURN) key. The following appears on the screen.

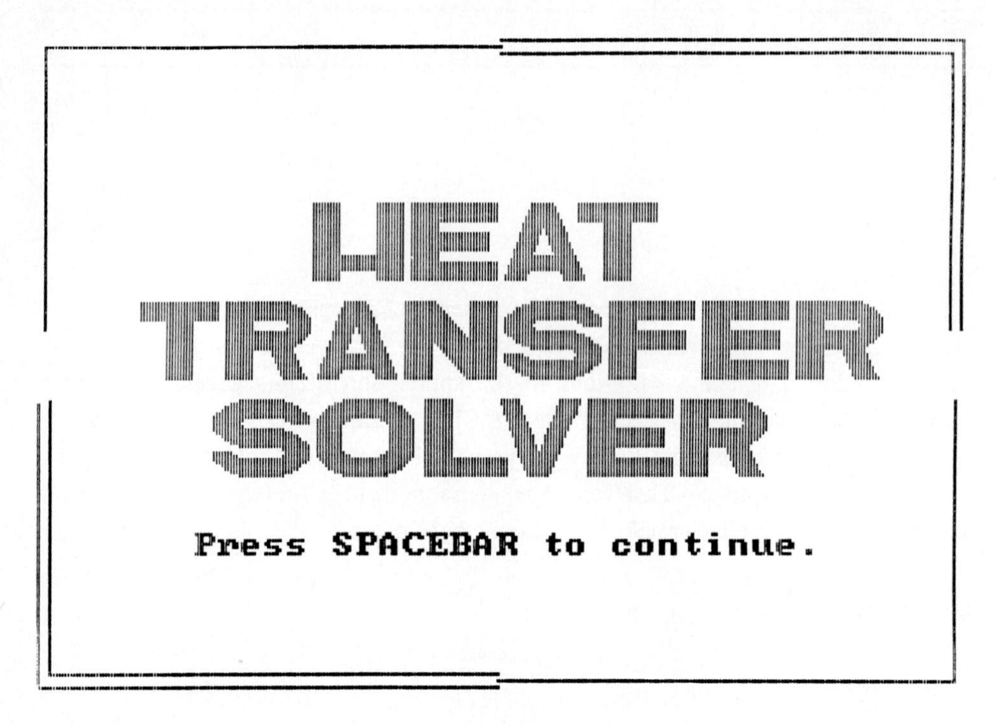

4. After pressing the SPACEBAR the screen displays

H E A T   T R A N S F E R   S O L V E R

M.D.MIKHAILOV
Applied Mathematics Center
Sofia, Bulgaria

M.N.OZISIK
North Carolina State University
Raleigh, North Carolina

All rights reserved. Prepared in the United States of America. No part of this software may be reproduced , stored in a retrieval system , or transmitted, in any form or by any means, electronic, mechanical , photocopying , recording , or otherwise, without the prior written permission of the authors.

Press SPACEBAR to continue.

5. Press SPACEBAR again to obtain the MAIN MENU

```
            H E A T        T R A N S F E R        S O L V E R
-------------------------------------------------------------------------
  STEADY STATE CONDUCTION                FILMWISE CONDENSATION

  COMPOSITE MEDIUM                       NUCLEATE BOILING

  LUMPED SYSTEM ANALYSIS                 FILM BOILING

  TRANSIENT CONDUCTION                   BLACKBODY RADIATION FUNCTION

  FINS                                   VIEW FACTORS

  FORCED CONVECTION                      RADIATION SHIELDS

  FREE CONVECTION                        HEAT EXCHANGER EFFECTIVENESS

F1Left F2Right F3Up F4Down F5Run task F7Colors                  <ESC>Exit
```

This MENU has two purposes: To select the desired program or desired colors for the background and characters.

PROGRAM SELECTION. Move the high-lighted indicater over any one of fourteen programs listed in the main menu by using the appropriate function key indicated at the bottom of the menu, i.e., *F1Left*, *F2Right*, *F3Up* and *F4Down*. Then select the desired program by pressing either the *F5Run task* function key or ENTER key. A brief description of each of these fourteen programs is obtainable by pressing "?". For example if the high-lighted indicator is on the COMPOSITE MEDIUM and ? key is pressed, the following HELP window appears on the screen.

```
            H E A T        T R A N S F E R        S O L V E R
-------------------------------------------------------------------------
  STEADY STATE CONDUCTION                FILMWISE CONDENSATION

  COMPOSITE MEDIUM                       NUCLEATE BOILING

  LUMPED SYSTEM ANALYSIS                 FILM BOILING

  TRANSIENT CONDUCTION           +----------------------------------+
                                 |                                  |
  FINS                           |   COMPOSITE  MEDIUM  program     |
                                 | calculates   one - dimensional   |
  FORCED CONVECTION              | steady-state  heat  flow  and    |
                                 | interface temperatures in slab,  |
  FREE CONVECTION                | hollow-cylinder, and hollow -    |
                                 | sphere of up to 9 layers.        |
                                 |                                  |
                                 | Press ANYKEY then procced as above.
F1Left F2Right F3Up F4Down F5Run task F
```

Press any key to return to the MAIN MENU.

COLOR SELECTION. A monochrome or a color screen can be used. When a color screen is used, the default colors for the background and characters are blue and white, respectively, for the main screen, and red and yellow for the HELP screen. To change the colors press *F7Colors* key as indicated at the bottom of the screen. The following appears

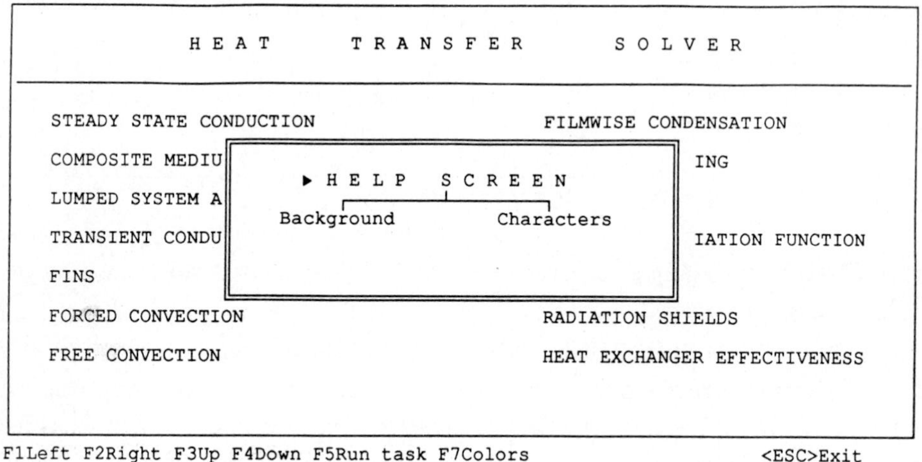

Blinking cursor appears before the MAIN SCREEN title. Each time the SPACEBAR is pressed the window alternately changes from MAIN SCREEN to HELP SCREEN, indicating the default colors for these screens. Suppose the screen displays the MAIN SCREEN window. By pressing the ENTER key, the cursor is moved before the *Background.* Now, each time the SPACEBAR is pressed, the color of the background changes. Press ENTER key when the desired color appears; then the cursor moves before the *Characters.* Each time the SPACEBAR is pressed the color of the characters changes. Press the ENTER key when the desired color appears; then the cursor moves before the MAIN SCREEN. To change to the HELP SCREEN window, press the SPACEBAR and then the ENTER key. The desired colors for the background and the characters of the HELP SCREEN can be changed in a similar manner described previously.

Once the colors are selected press ESC key to return to the MAIN MENU.

ENTERING THE INPUT DATA

The general procedure for entering the INPUT data into the program is now described. However, there are differences in specific applications for materials covered in different chapters. Therefore, the reader should go over the first example in each chapter for the explanation of specific details.

Suppose the screen displays the MAIN MENU and we wish to select the program FORCED CONVECTION. Pressing F4Down function key move the high-lighted indicator over the FORCED CONVECTION and then press the ENTER key. The screen displays the following

```
                FORCED CONVECTION

          Type of flow : Inside ducts
```

The *high-lighted indicator* appears over a built-in quantity *Inside ducts*. Each time the SPACEBAR is pressed the type of flow successively changes to *Inside ducts* and *Over bodies*. When the desired type of flow appears, say, *Inside duct*, press the ENTER key. The screen changes to

```
                FORCED CONVECTION

          Type of flow : Inside ducts
              Geometry : Circular tube
```

Now the high lighted indicator is over the Circular tube. Each time the SPACEBAR is pressed, the INPUT alternates between *Parallel plate channel* and *Circular tube*. Press the ENTER key when *Circular tube* appears. Then the screen displays the following

```
                          FORCED CONVECTION

                   Type of flow : Inside ducts
                       Geometry : Circular tube

                       Diameter : m                    0.000
                         Length : m                    0.000
             Mean flow velocity : m/s                  0.000
     Fluid thermal conductivity : W/m•K                0.00000000
      Fluid kinematic viscosity : m²/s                 0.00000000


                Prandtl number :                       0.000

               Reynolds number =                          0.000
     Heat transfer coefficient = W/m²•K                   0.000
```

The next item, Diameter, has built-in units m, cm, mm, in, ft. To select the unit for the diameter, press the SPACEBAR until the desired unit is displayed. Press the ENTER key when cm appears. The cursor moves immediately to the right, starts blinking and is ready to accept the numerical value of the diameter. Enter the value 2.5, press the SPACEBAR and then the ENTER key. Suppose a value of zero, which is unacceptable, is entered for the tube diameter. As soon as the ENTER key is pressed, blinking red light is displayed over the unacceptable value. In such a case, use the BACKSPACE or DEL key to clear, enter the correct value, press the SPACEBAR and then the ENTER key.

While entering the INPUT data, if the NEW DATA touches the OLD DATA which was already on the screen, press the SPACEBAR to separate them, and then press the ENTER key. Otherwise a blinking red light will appear over the data when the ENTER key is pressed. In such a case, use BACKSPACE or DEL key to delete the entry, then enter the NEW DATA and press the ENTER key.

The units and the values of the *Length, Mean flow velocity, Fluid thermal conductivity, Fluid viscosity* and the *Prandtl number* are selected and entered in a similar manner. The following figure shows a typical selection for each of these items.

```
                    FORCED CONVECTION

            Type of flow : Inside ducts
                Geometry : Circular tube

                Diameter : cm                    2.500
                  Length : m                    40.000
      Mean flow velocity : m/s                   0.700
Fluid thermal conductivity : W/m•K               0.14400000
 Fluid kinematic viscosity : m²/s                0.00024000


          Prandtl number :                    2870.000

         Reynolds number =                        0.000
 Heat transfer coefficient = W/m²•K               0.000
```

To run the program, press the F5 function key. Then the screen displays the following

```
                    FORCED CONVECTION

            Type of flow : Inside ducts
                Geometry : Circular tube

                Diameter : cm                    2.500
                  Length : m                    40.000
      Mean flow velocity : m/s                   0.700
Fluid thermal conductivity : W/m•K               0.14400000
 Fluid kinematic viscosity : m²/s                0.00024000


          Prandtl number :                    2870.000

         Reynolds number =                       72.917
 Heat transfer coefficient = W/m²•K              44.913
```

Note that, the OUTPUT data *Reynolds number* and the *H. T. coefficient* have equal sign in front of them, whereas all the INPUT data have colons in front of them.

A special feature of this software is that the units and/or the values of the INPUT data can be easily changed by moving the cursor over the desired quantity, entering the new units and values, and then running the program by pressing the F5 function key. Another feature is that, the screen will display a warning message whenever the correlation limits are violated as a result of improper input DATA. To illustrate both of these features we change the diameter from 2.5 cm to 1 m and run the program. The screen displays the warning sign *Outside the correlation range* as shown in the following Table.

```
                         FORCED CONVECTION

              Type of flow : Inside ducts
                  Geometry : Circular tube

                  Diameter : m                       1.000
                    Length : m                      40.000
        Mean flow velocity : m/s                     0.700
Fluid thermal conductivity : W/m•K                   0.14400000
 Fluid kinematic viscosity : m²/s                    0.00024000


           Prandtl number :                       2870.000

          Reynolds number =                          0.000
 Heat transfer coefficient = W/m²•K                  17.133

Outside the correlation range
```

GRAPHICAL SUPPORT

The results can be represented in the graphical form by pressing F8(*Graphic*) function key. The screen displays the coordinate axes *H. T. coefficient vs Flow velocity* and automatically sets the units and the ranges of variables as shown below. Press F8(*Graphic*) once more, the screen will display "please wait! calculating". Soon afterwards the graphical representation of the results will appear on the screen. To return to the Table, first press F8 and then ESC. The user can edit the units as well as the end values of the *H. T. Coefficient* and *Flow velocity* by moving the cursor using F1, F2, F3 or F4 function keys. The unit is changed simply by pressing the SPACEBAR when the cursor is over the unit. The values are changed by entering the desired value when the cursor is over the number to be replaced and then pressing the RETURN key. In this

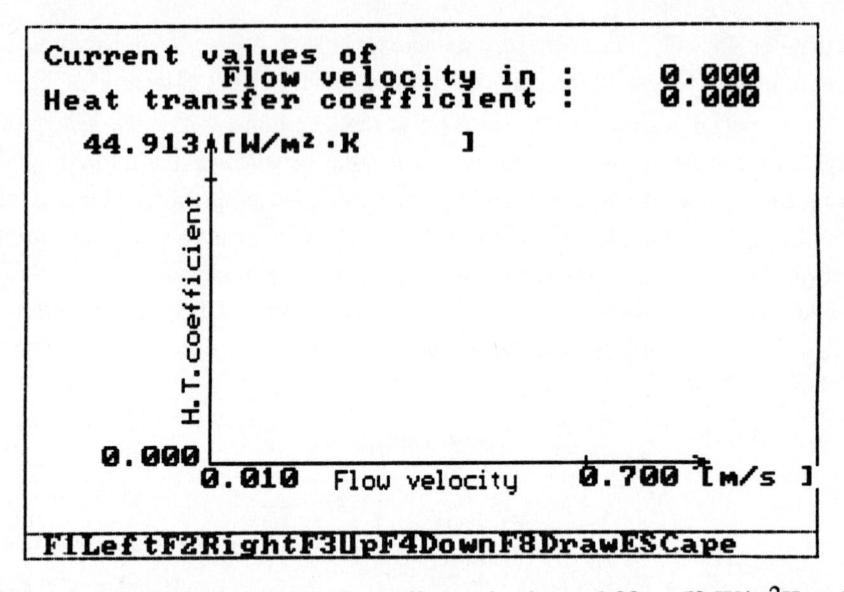

example, the end values of the *H. T. coefficient* is changed 20 to 50 W/m^2K and the program is run by pressing the F8 Draw function key. The following graphical representation appears on the screen.

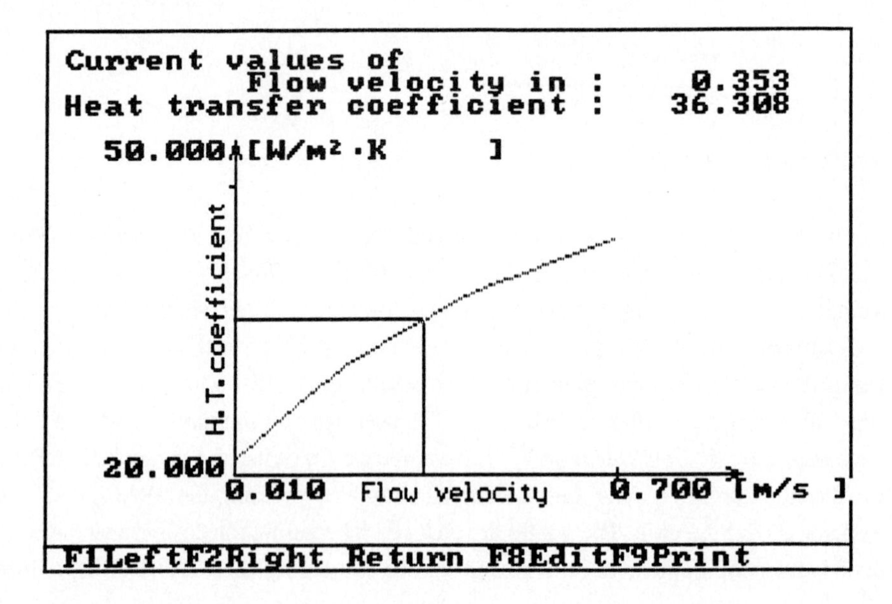

To escape from the graphic mode, first press F8(Edit) function key and then the ESC key; "Forced Convection" data entry table will appear on the screen.

TABULATION OF RESULTS

In the graphical form each time the RETURN key is pressed, the *H. T. coefficient* and *Flow velocity* corresponding to the cursor location are stored in the memory of the computer. Up to 150 such values can be stored. To obtain a hard copy of the stored results press F9 function key a typical listing of stored results is shown in the following table.

FORCED CONVECTION

Type of flow : Inside ducts
Geometry : Circular tube

Velocity [m/s]	H.T. coeficient [W/m²•K]
0.200	31.149
0.302	34.813
0.404	38.082
0.501	40.556
0.561	41.855

HEAT TRANSFER SOLVER

Chapter

ONE

STEADY STATE CONDUCTION

This chapter is concerned with the computer solution of one-dimensional steady-state heat conduction in a slab, hollow cylinder and hollow sphere subjected to prescribed constant temperatures at both boundary surfaces. We assume no energy generation in the medium. The analysis is developed for the case of constant thermal conductivity; however, the program can handle temperature dependent thermal conductivity by using the effective thermal conductivity.

The problems involving prescribed surface heat flux and convection boundary conditions are considered in the next chapter as a special case of the composite medium problems.

We present below, first, an analysis of the problem, and then illustrate the use of the computer solutions with representative examples.

1-1 ANALYSIS

Consider one-dimensional steady-state heat flow through bodies such as a slab, hollow cylinder and hollow sphere as illustrated in Fig. 1-1. The inner and outer surfaces are kept at constant temperatures T_a and T_b, respectively.

The total heat flow rate Q across the slab, hollow cylinder and hollow sphere is expressed in the form

$$Q = \frac{T_a - T_b}{R} \tag{1-1}$$

where T_a and T_b are the inner and outer surface temperatures, respectively. The thermal resistance R is given by

$$R = \frac{L}{kA} \qquad \text{for slab} \tag{1-2a}$$

$$R = \ln\left(\frac{d+2L}{d}\right)/(2\pi kH) \qquad \text{for cylinder} \tag{1-2b}$$

$$R = \left(\frac{1}{d} - \frac{1}{d+2L}\right)/(2\pi k) \qquad \text{for sphere} \tag{1-2c}$$

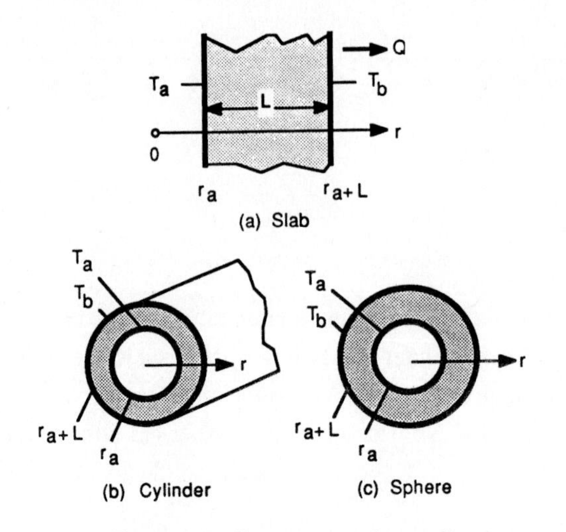

Figure 1-1 Geometry and coordinates

where

L = thickness of the slab, hollow cylinder or hollow sphere
k = thermal conductivity
A = area of the slab
d = inside diameter of cylinder or sphere
H = length of cylinder.

The temperature distribution T across the layer is given as a function of the distance x measured from the inner surface and related to the r coordinate by $x=4-r_a$. Then

$$T(x) = T_a + (T_b - T_a)\frac{x}{L} \qquad \text{for slab} \qquad (1\text{-}3a)$$

$$T(x) = T_a + (T_b - T_a)\ln\left(\frac{d}{d+2x}\right) \Big/ \ln\left(\frac{d}{d+2L}\right) \qquad \text{for cylinder} \qquad (1\text{-}3b)$$

$$T(x) = T_a + (T_b - T_a)\frac{\frac{1}{d} - \frac{1}{d+2x}}{\frac{1}{d} - \frac{1}{d+2L}} \qquad \text{for sphere} \qquad (1\text{-}3c)$$

where $x=r-r_a$

If the thermal conductivity k(T) depends on temperature, then, the *effective thermal conductivity*, defined as

$$k_{eff} = \frac{1}{T_a - T_b}\int_{T_b}^{T_a} k(T)\, dT$$

has to be used in Eqs. (1-2) to replace k.

1-2 COMPUTER SOLUTIONS

The program *STEADY STATE CONDUCTION* performs one-dimensional steady-state heat conduction calculations for plate, hollow cylinder, and hollow sphere by using the analytic expressions described previously. A unique feature applicable only to this *STEADY STATE CONDUCTION* program is that, any one of the parameters involved in the calculations can be OUTPUT while all the others remain as INPUT. The INPUT data always appear after the colon sign ":", while the OUTPUT data after the equal sign "=". It is only in this program that, each time the SPACEBAR is pressed when the cursor is over the ":" or "=" sign, the sign changes alternately from ":" to "=". The ENTER key is pressed when the desired sign appears on the screen. Suppose the equal sign "=" is chosen; then the quantity associated with this sign becomes the OUTPUT.

The GRAPHIC support is also available for the program. That is, the temperature distribution in the medium can be plotted as a function of the distance. In addition, the exact values of "temperature and distance" corresponding to the cursor location are displayed at the upper right-hand side of the graph. If the cursor is moved along the graph, the current values of "temperature and distance" are continuously displayed. Each time the ENTER key is pressed, the current values of the "temperature and distance" are registered in the memory of the computer and these results can be printed by pressing the F9 function key.

The use of the program Steady State Conduction is now illustrated with representative examples. In the first example a detailed description is given of entering the INPUT data and using the GRAPHICAL support.

Example 1-1. A temperature difference of 200C is applied across a corkboard 5 cm thick, having thermal conductivity 0.04 W/m ·C as illustrated in the figure below. Calculate the heat transfer rate across 3 m^2 of the board per hour.

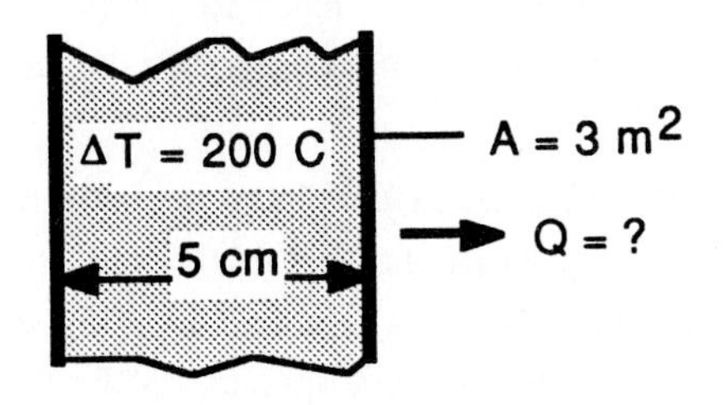

Solution. Table below shows the computer INPUT and OUTPUT for the Example 1-1 after the program has been run. Note that only *Heat flow rate* has equal sign in front of it, because it is the OUTPUT data, and all other quantities have column sign in front of them, because they are INPUT data. We now describe the main features of entering the INPUT data and the use of the GRAPHICAL support for the problem.

```
                    STEADY  STATE  CONDUCTION

                     Geometry : Plate

                         Area : m²                         3.000

                    Thickness : cm                         5.000
         Thermal conductivity : W/m•K                      0.040
    Inner surface temperature : C                        200.000
    Outer surface temperature : C                          0.000
               Heat flow rate = W                        480.000
```

Entering the INPUT data. The first INPUT data is the geometry (i.e., slab, cylinder or sphere). To set the geometry, move the cursor next to the *Geometry* and press the SPACEBAR. Each time the SPACEBAR is pressed, the type of geometry changes, successively, to *Plate, Cylinder, Sphere.* In this example the geometry is plate. Therefore, when *Plate* appears on the screen, press the ENTER key in order to set the geometry to *Plate.*

The next item is the *Area* which is an INPUT data for this problem; therefore the sign in front of it should be set to column ":". Note that, when the cursor is over the sign, each time the SPACEBAR is pressed the sign is alternately changes from ":" to "=". Press the ENTER key when the desired sign appears on the screen. After choosing the sign, the cursor moves over m^2 which is the UNIT for the area. Each time the SPACEBAR is pressed the units change successively to cm^2, mm^2, ft^2, in^2, m^2, and so on. In this example the unit for the area is in m^2; therefore when m^2 appears on the screen press the RETURN key in order to set the unit to m^2. Immediately afterwards, the cursor moves to the right and starts blinking, ready to accept the numerical value of the area. For this problem the area is 3 m^2. Therefore, enter the value 3, and then press the ENTER key.

The next item is the *Thickness* of the plate. In this example the thickness is an INPUT data, therefore set the sign in front of the thickness to column ":" and then press the ENTER key. The thickness is 5 cm. Therefore, when the cursor is over the UNIT, press the SPACEBAR successively until cm appears on the screen and then press the ENTER key. The cursor moves to the right, starts blinking and is ready to accept the numerical value of the thickness which is 5. Type 5 and then the ENTER key.

The next item is the *Thermal conductivity* which is an INPUT data therefore the column sign ":" should be in front of it. Move the cursor over the unit. Each time the SPACEBAR is pressed the unit changes successively, to W/cm·K, Btu/ft·h·F, kcal/m·h·K, W/m·K and so on. Set the unit to W/m·K, and then enter its numerical value 0.04.

The next two items are the *Inner surface temperature* and *Outer surface temperature*, respectively, both of which are the INPUT data. In the present problem the temperature difference is specified as 200C. Therefore, the values assigned to the surfaces should be such that their difference should be 200C. We have chosen the *Inner surface temperature* to be 200C and the *Outer surface temperature* 0C. As discussed previously, the signs are set to ":" because they are INPUT data, the units for temperature are set to C. Enter the value 200 for the inner surface temperature. This value happens to touch the previous value displayed on the screen. They should be separated from touching each other by pressing the SPACEBAR. Therefore, after typing 200, press the SPACEBAR and then the ENTER key.

Finally enter the value 0 for the outer surface temperature.

The computer is now ready to RUN the program. Press the F5 function key to run the program. The OUTPUT gives

Heat Flow Rate = 480.0 W.

By pressing the SPACEBAR the answer can be expressed in different units including kW, cal/s, kcal/h, Btu/h and W.

Any of the INPUT data can be changed by moving the cursor to the appropriate input data, entering the new value and running the program again.

Limits are imposed on the acceptable value of each INPUT data. For example the thickness of the slab is restricted to $0<L<9999$ m. If such a limit is exceeded, blinking will occur over the unacceptable input data. Blinking will stop when correct data is entered.

Graphical Support. The results can also be presented in the graphical form if the graphical support is activated by pressing the F8 function key. The screen displays the coordinate axes "Temperature vs. Distance", automatically sets the units, and the maximum ranges for the temperature and distance based on the values specified in the example, i.e., $0 \leq T \leq 200$C and $0 \leq x \leq 5$ cm. The user can edit the units as well as the end values of these intervals by moving the cursor using F1, F2, F3 or F4 function keys. The units are changed simply by pressing the space bar when the cursor is over the unit. The values are changed by entering the desired number when the cursor is over the number to be replaced and then pressing the return key. One can narrow the range of the distance or temperature by altering the numerical values of the range for distance and temperature on the graph. This operation is equivalent to zooming to a specified portion of the graph. If the edited values of temperature or distance on the graph exceeds the internally set maximum permissible ranges, then the curve will not be plotted when F8 function key is pressed. In such a case revise the end values on the graph.

To draw the curve press F8 function key; the following figure will appear on the screen. By pressing the F8 key successively, one can change from the "F8 Edit" to "F8 Draw" modes.

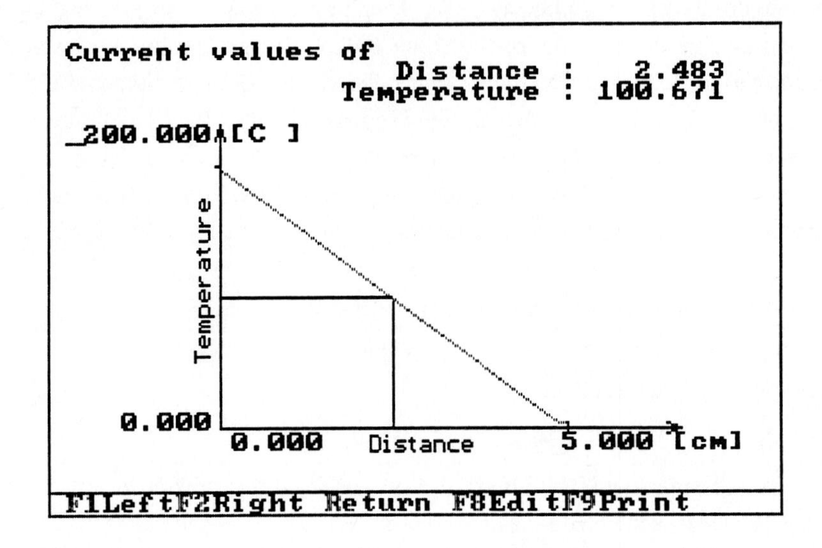

After the graph is plotted, if further changes are needed in the units chosen or in the values of the ranges of the variables, or Escape is needed, press the F8 function key in order to change into the Edit mode.

STEADY STATE CONDUCTION

Geometry : Plate

Distance [cm]	Temperature [C]
0.034	198.658
0.805	167.785
1.510	139.597
2.081	116.779
2.483	100.671
2.987	80.537
3.490	60.403
3.993	40.268
4.463	21.477

Tabulation of the Results. In the graphical form, each time the ENTER key is pressed, the temperature and distance corresponding to the cursor location are stored in the memory of the computer. Up to 150 such values can be stored, since the computer performs the calculations at 150 equally divided locations over the space interval considered.

To obtain a hard copy of the results stored in the memory, press the F9 function key. The stored results are printed as shown in the previous Table.

Example 1-2. Glass wool of thermal conductivity k = 0.038 W/m·K is to be used to insulate an ice box. If the maximum heat loss should not exceed 45 W/m^2 for a temperature difference of 40C across the walls of the refrigerator, determine the thickness of the insulation. Figure below shows the geometry.

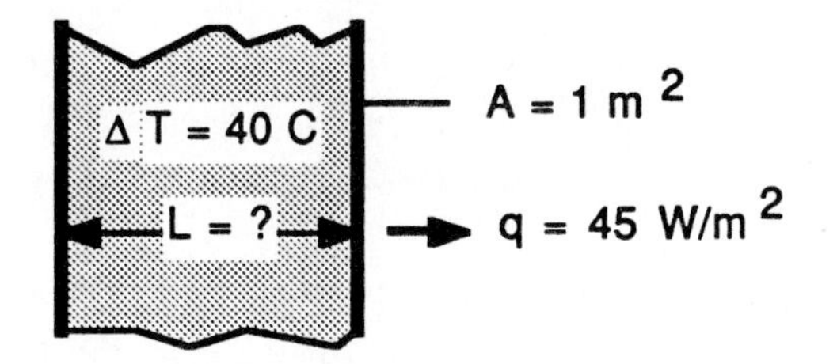

Solution. Table below shows the computer INPUT and OUTPUT data for this problem. Note that, instead of specifying the inside and outside surface temperatures T_a and T_b separately, only the temperature difference ΔT = 40C between the surfaces is specified. The steady-state heat flow is invariant of the magnitudes of the temperatures T_a and T_b so long as their difference T_a and T_b remain equal to the specified value of ΔT = 40C. Therefore, in this example, for convenience, the INPUT data for the surface temperatures are chosen as T_a = 40C and T_b = 0C.

The OUTPUT data for this problem is the plate thickness L; therefore, when entering the INPUT data, the sign in front of the *Thickness* should be changed to the equal sign "=". The area is taken as 1 m^2, because the heat flow rate is for one meter

```
                STEADY  STATE  CONDUCTION

                 Geometry : Plate

                     Area : m²                    1.000

                Thickness = cm                    3.378
     Thermal conductivity : W/m•K                 0.038
Inner surface temperature : C                    40.000
Outer surface temperature : C                     0.000
           Heat flow rate : W                    45.000
```

square. The solution gives the plate thickness L = 3.378 cm. If division by zero occurs during computation, a blinking sign will appear on the screen, warning that the input data are incorrect. In such a case press the SPACEBAR and change the input data. For example, in the above Table, if the *Heat flow rate* is changed from 45 to 0, and the program is run by pressing the F5 function key, such a warning sign will appear.

Example 1-3. The heat flow rate across an insulating plate of thickness 3 cm, thermal conductivity 0.1 W/m·K is 250 W/m^2. If the hot surface temperature is 175C, what is the temperature of the cold surface? Figure below shows the geometry.

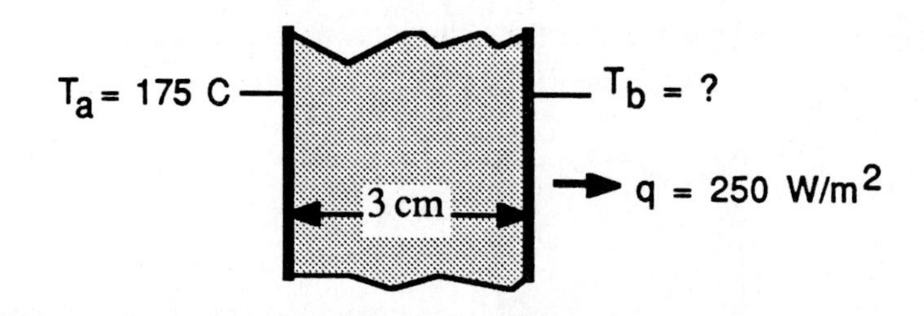

Solution. Table below shows the INPUT and OUTPUT data for this problem. The outer surface temperature T_b is unknown, hence it is the OUTPUT data and all the remaining quantities are the INPUT data.

STEADY STATE CONDUCTION

Geometry : Plate

Area :	m²	1.000
Thickness :	cm	3.000
Thermal conductivity :	W/m•K	0.100
Inner surface temperature :	C	175.000
Outer surface temperature =	C	100.000
Heat flow rate :	W	250.000

Example 1-4. The heat flow rate through a 4-cm thick wood board, for a temperature difference of 25C between the boundary surfaces is 75 W/m^2 as illustrated in the figure below. What is the thermal conductivity of the wood?

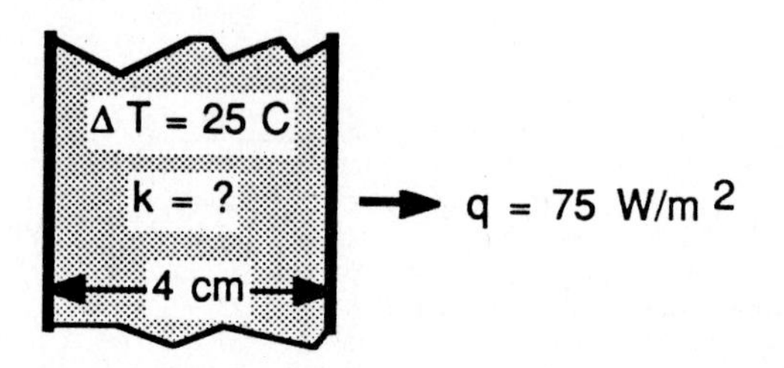

Solution Table below shows the computer INPUT and OUTPUT data for this example. As the temperature difference between the surfaces, $T_a - T_b = \Delta T = 25C$ is specified, the INPUT data for the surface temperatures are chosen, for convenience, as $T_a = 25C$ and $T_b = 0C$. The unknown or the OUTPUT for this problem is the thermal conductivity. Therefore, entering the INPUT data the sign in front of Thermal Conductivity has been changed from the column ":" to the equal sign "=". As discussed previously, the sign change is made by moving the cursor above the column sign ":" in front of the thermal conductivity. Then, press the SPACEBAR to change ":" to "=" and press the ENTER key to set the sign to the equal sign.

The solution gives the thermal conductivity k = 0.120 W/m·K as the OUTPUT for the problem.

```
                STEADY  STATE  CONDUCTION

                  Geometry : Plate

                      Area : m²                     1.000

                 Thickness : cm                     4.000
      Thermal conductivity = W/m•K                  0.120
 Inner surface temperature : C                     25.000
 Outer surface temperature : C                      0.000
           Heat flow rate : W                      75.000
```

Example 1-5. A cylindrical insulation for a steam pipe has an inside radius $r_a = 6$ cm, outside radius $r_b = 8$ cm and a thermal conductivity k = 0.5 W/m·C. The inside surface of the insulation is at a temperature $T_a = 430C$ and the outside surface at $T_b = 30C$ as illustrated in the figure below. Determine the heat loss per meter length of the insulation.

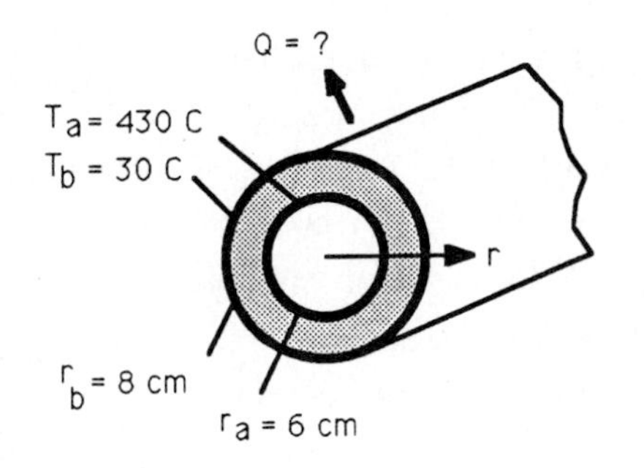

Solution. Table below shows the INPUT and OUTPUT data for this problem. Note that the Heat Flow Rate being unknown it is the output data for this problem, therefore it has the equal sign "=" in front of it.

STEADY STATE CONDUCTION

Geometry : Cylinder

Length	:	m	1.000
Inner surface radius	:	cm	6.000
Thickness	:	cm	2.000
Thermal conductivity	:	W/m•K	0.500
Inner surface temperature	:	C	430.000
Outer surface temperature	:	C	30.000
Heat flow rate	=	W	4368.145

Starting from the top of this Table, the geometry is entered as *Cylinder*. This is done by moving the cursor over the type of geometry and successively pressing the SPACEBAR until *Cylinder* appears on the screen. Then the ENTER key is pressed to set the geometry to *Cylinder*.

The next item is the *Length* of the cylinder. For this particular example it is H = 1 m. Therefore, first the UNIT is set to m and then the value 1 is entered.

Then, the inner surface radius r_a=6 cm and the thickness $L=r_b-r_a$=2 cm are entered.

The values of the thermal conductivity, the inner and outer surface temperatures are then entered.

Finally, the last item Heat Flow Rate is the unknown quantity, hence it is the OUTPUT for the problem. Only the unit of this quantity is chosen; the corresponding value appears on the screen after the program has been RUN by pressing the F5 function key.

For the problem considered here the OUTPUT is Q=4368.145W. This output can be expressed in other units, by merely moving the cursor over the unit for the heat flow rate and pressing the SPACEBAR. Each time the SPACEBAR is pressed, the unit changes successively to W, kW, cal/s, kcal/h and Btu/hr, and the computer immediately adjusts the numerical value of Q to the chosen new units.

Example 1-6. Determine the heat flow rate through a spherical copper shell of thermal conductivity k=386 W/m·K, inner radius r_b=2 cm and outer radius r_b=6 cm if the inner surface is kept at T_a=200C and the outer surface at T_b=100C. Figure below shows the geometry.

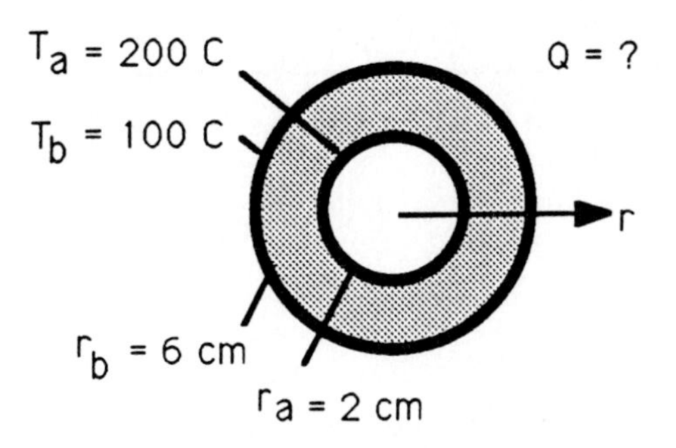

Solution. The computer INPUT and OUTPUT data are shown in Table below. The INPUT data are entered as discussed for the cylinder. The Heat Flow Rate is the OUTPUT for the problem and its value is determined after the program has been run as Q = 14551.857W.

STEADY STATE CONDUCTION

Geometry : Sphere

Inner surface radius	:	cm	2.000
Thickness	:	cm	4.000
Thermal conductivity	:	W/m•K	386.000
Inner surface temperature	:	C	200.000
Outer surface temperature	:	C	100.000
Heat flow rate	=	W	14551.857

Chapter

TWO

COMPOSITE MEDIUM

In many engineering applications, heat transfer takes place through a medium composed of several layers, each having different thermal conductivity. Consider, for example, a hot fluid flowing inside a tube covered with a uniform layer of thermal insulation. The thermal conductivities of the tube metal and insulation are different; hence the heat flow from the hot fluid to the colder outer environment takes place through a composite medium consisting of two parallel concentric cylinders. In this chapter we examine, the determination of heat flow rate and the interface temperatures for one-dimensional steady-state heat conduction through a composite medium consisting of several layers of slabs, coaxial cylinders and concentric spheres.

2.1 ANALYSIS

Figure below illustrates a composite structure consisting of several layers of slabs, coaxial cylinders or concentric spheres. Thermal conductivity of each layer is independent

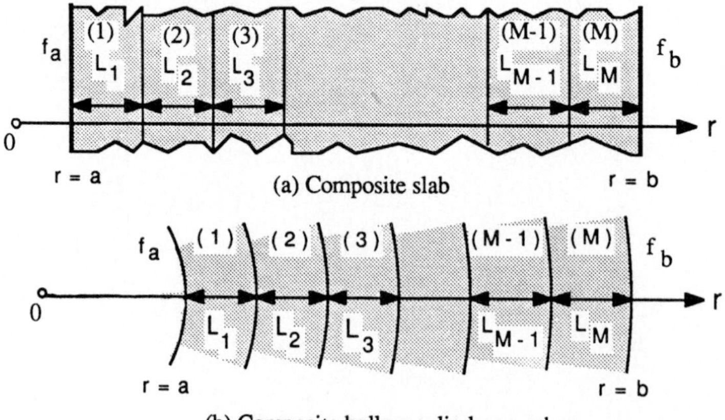

(a) Composite slab

(b) Composite hollow cylinder or sphere

of temperature, but constant within the layer and different for each layer. There is no energy generation in the medium. The boundary conditions for the outer surfaces at r=a and r=b may be prescribed temperature, prescribed heat flux or convection with a specified heat transfer coefficient into a surrounding at a prescribed constant temperature. We assume perfect thermal contact between the layers. The *thermal resistance concept* can be used to determine the heat flow rate $Q^{(n)}$ and the interface temperatures. The resulting expression for the heat flow rate $Q^{(n)}$ is obtained as a special case from the general expression given by Mikhailov and Özisik(1984, p. 198)

$$Q^{(n)} = \frac{(\alpha_b f_a - \alpha_a f_b) A_n}{\frac{\alpha_b \beta_a}{a^n} + \frac{\alpha_a \beta_b}{b^n} + \alpha_a \alpha_b \sum_{i=1}^{M} R_i^{(n)}} \qquad (2\text{-}1)$$

where

$$n = \begin{cases} 0 & \text{for slab} \\ 1 & \text{for cylinder} \\ 2 & \text{for sphere} \end{cases}$$

M = number of layers

$Q^{(n)}$ = total heat transfer rate through the composite structure

The choice of the parameter α_a, β_a, f_a depends on the type of the boundary condition at the surface r=a. Similarly, the choice of the parameters α_a, β_b, f_b depends on the type of boundary conditions at the boundary surface r=b. Here, the following three different types of boundary conditions are considered: prescribed surface temperature, prescribed surface heat flux, and convection into a surrounding at a prescribed constant temperature with a given heat transfer coefficient.

Clearly, there are nine different combinations of boundary conditions. In the case of prescribed heat flux at both boundaries, r=a and r=b, the problem has no unique solution. Therefore, the computer will not accept a choice of prescribed heat flux at both surfaces. Depending on the type of the boundary condition at the boundary surfaces r=a and r=b, the values of the parameters α_a, β_a, f_a and α_b, β_b, f_b are chosen as listed in the Table 2-1. In this table, the parameters h_a and h_b are the heat transfer coefficients at the boundary surfaces r=a and r=b, respectively.

Table 2-1 Values of the parameters α_a, β_a, f_a and α_b, β_b, f_b.

Boundary Surface at r=a		
α_a	β_a	f_a
1	0	value of prescribed surface temperature
0	1	value of prescribed surface heat flux
1	$1/h_a$	value of prescribed surrounding temperature

Boundary Surface at r=b		
α_b	β_b	f_b
1	0	value of prescribed surface temperature
0	1	value of prescribed surface heat flux
1	$1/h_b$	value of prescribed surrounding temperature

In Eq. (2-1), the area parameter A_n, (n=0,1,2), depends on the geometry; for a slab, hollow cylinder and hollow sphere it is given by:

Slab (n=0): $$A_0 \equiv A = \text{surface area} \tag{2-2a}$$

Cylinder (n=1): $$A_1 = 2\pi H \tag{2-2b}$$

where H is the cylinder length, and

Sphere (n=2): $$A_2 = 4\pi. \tag{2-2c}$$

Finally, in Eq. (2-1), the thermal resistance parameter $R_i^{(n)}$, for any layer-i, depends on the geometry; for a slab, hollow cylinder and hollow sphere, it is given by:

Slab (n=0): $$R_i^{(0)} = L_i/k_i \tag{2-3a}$$

where L_i is the thickness and k_i is the thermal conductivity of the layer-i,

Cylinder (n=1): $$R_i^{(1)} = \ln(r_i/r_{i-1})/k_i \tag{2-3b}$$

where r_{i-1} and r_i are the inner and outer radius, respectively and k_i is the thermal conductivity of the layer-i,

Sphere (n=2) $$R_i^{(2)} = \left(\frac{1}{r_{i-1}} - \frac{1}{r_i}\right) \Big/ k_i \tag{2-3c}$$

where r_{i-1} and r_i are the inner and outer radius, respectively, and k_i is the thermal conductivity of the layer-i.

where r_{i-1} and r_i are the inner and outer radius, respectively, and k_i is the thermal conductivity of the layer-i.

The foregoing equations are programmed to compute the heat transfer rate $Q^{(n)}$ through the composite layer. The thermal resistance concept is used to determine the interface temperatures.

2-2 CONTACT CONDUCTANCE AT INTERFACES

The above analysis assumes perfect thermal contact between the layers. If there is thermal contact conductance or resistance between the layers, the effects of contact resistance to heat flow rate can be included in the calculations by treating the contact resistance as an equivalent *fictitious layer* of negligible thickness having a thermal resistance equal to that of the contact resistance between the layers.

Suppose we have a contact conductance h_i between the layers (i-1) and (i). This can be regarded as a thermal resistance $R_i^{(n)}$ as

$$R_i^{(n)} \equiv \frac{1}{h_i} \tag{2-4}$$

where

$$n = \begin{cases} 0 & \text{for slab} \\ 1 & \text{for cylinder} \\ 2 & \text{for sphere} \end{cases}$$

Then, the thermal resistance $R_i^{(n)}$ defined by Eq. (2-4) is introduced into the summation term appearing in the denominator of Eq. (2-1). To represent $R_i^{(n)}$ as a fictitious layer, the thermal conductivity and thickness of the fictitious layer are so chosen that the resulting thermal resistance of the layer should be equal to $R_i^{(n)}$. The procedure for determining the thickness and thermal conductivity of a fictitious layer to satisfy a given h_i or $R_i^{(n)}$ for a slab, cylinder and sphere is as follows.

slab (n=0):

$$R_i^{(0)} \equiv \frac{L_i}{k_i} = \frac{1}{h_i} \tag{2-5}$$

Choose a value for L_i and then determine k_i so that Eq. (2-5) is satisfied for the given value of h_i.

Cylinder (n=1):

$$R_i^{(1)} \equiv \frac{\ln(r_i/r_{i-1})}{k_i} = \frac{1}{h_i} \tag{2-6}$$

Choose a value for the thickness of the fictitious layer and compute the ratio r_i/r_{i-1}. Then determine k_i so that Eq. (2-6) is satisfied for the specified value of h_i.

Sphere (n=2): $$R_i^{(2)} \equiv \left(\frac{1}{r_{i-1}} - \frac{1}{r_i}\right)\frac{1}{k_i} = \frac{1}{h_i} \quad (2\text{-}7)$$

Choose a value for the thickness of the fictitious layer and compute the quantity $\left(\frac{1}{r_{i-1}} - \frac{1}{r_i}\right)$. Then determine k_i so that Eq. (2-7) is satisfied for the given value of h_i.

It is to be noted that, for the case of the slab, any value can be chosen for the thickness L_i of the fictitious layer. However, for the case of cylinder and sphere, the thickness L_i of the fictitious layer should be so chosen as $L_i << r_{i-1}$ in order not to disturb the original geometry.

2-3 COMPUTER SOLUTIONS

The analytic expressions presented previously are programmed to compute steady state heat transfer rate $Q^{(n)}$ (n=0,1,2) through a *composite slab, hollow cylinder* or *hollow sphere* consisting of up to nine layers, all in perfect thermal contact. The contact conductance (or contact resistance) between any two layers can be included in the calculations by treating the contact resistance as a fictitious layer as discussed previously. The program also calculates the interface temperatures and the temperatures at the outer boundary surfaces.

The graphic support is also available for the program. The temperature distribution at the layers can be plotted as a function of the position. In addition, the exact values of "temperature and position" corresponding to the cursor location are displayed at the upper right-hand side of the graph. The values of "temperature and position" up to 150 locations over the considered interval can be stored in the memory of the computer and printed by the printer.

The use of the program COMPOSITE MEDIUM is now illustrated with representative examples. Only in the first example a detailed description is given of entering the INPUT data and utilizing the GRAPHICAL support.

Example 2-1. The wall of an industrial furnace, illustrated in figure below consists of a fire clay brick of thickness L_1=0.2 m, thermal conductivity k_1=1 W/m·K and an insulation of thickness L_2=0.03 m, thermal conductivity k_2=0.05 W/m·K. The two layers are considered in perfect thermal contact (i.e., negligible contact resistance between the layers). The inside and outside surfaces of the wall are kept at 830C and 30C, respectively. Calculate the heat transfer rate across the furnace wall per meter square surface and the interface temperature $T_{1,2}$ between the wall and the insulation.

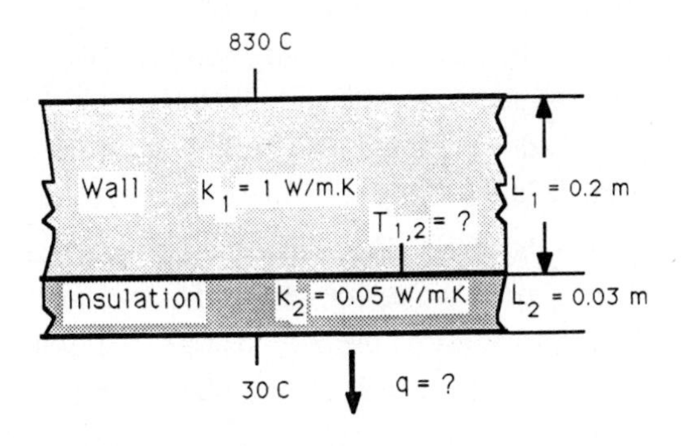

Solution. We present below the main features of entering the INPUT data, the use of the GRAPHICAL support and obtaining the hard copy of the results.

Entering the INPUT data. Table below shows the computer INPUT and OUTPUT data for this program. Note that the INPUT data always are followed by the colon sign ":", whereas the OUTPUT data are followed by the equal sign "=". However, in this program the INPUT and OUTPUT data cannot be interchanged as it has been the case for the program *STEADY STATE CONDUCTION*. Here the OUTPUT data are the *Heat flow rate* and the *Temperatures* of the interfaces and the outer surfaces. Clearly, for the prescribed heat flux and convection boundary conditions the outer surface temperatures are unknown; for the prescribed temperature boundary conditions they coincide with the specified surface temperatures.

```
                 COMPOSITE MEDIUM
                 Geometry : Plate
                     Area : m²                          1.000
      Surface temperature : C                         830.000


          Number of layers : 2
Layer thickness           Thermal conductivity          Temperatures
m          0.200     W/m•K              1.000         C        830.000
m          0.030     W/m•K              0.050         C        630.000
                                                      C         30.000



      Surface temperature : C                          30.000

            Heat flow rate = W                       1000.000
```

The first INPUT data is the setting up of the type of *Geometry*. To set the geometry, move the cursor next to the *Geometry* and press the SPACEBAR. Each time the SPACEBAR is pressed, the type of geometry changes, successively, to *Plate*, hollow *Cylinder* and hollow *Sphere*. Press the ENTER key when the appropriate geometry is displayed. In this example it is a *Plate*.

The next item is the *Area*. First enter the unit and then the magnitude of the area. To specify the unit, move the cursor next to the *Area* and press the SPACEBAR. Each time the SPACEBAR is pressed the unit changes, successively, to m^2, cm^2, mm^2, ft^2 and in^2. In this example the area is in m^2. Therefore, when m^2 appears on the screen, press the ENTER key to set the unit to m^2. Then the cursor immediately moves to the right, starts blinking and is ready to accept the numerical value of the area. In this example the area is 1 m^2; therefore enter the value of area as 1.

The next item is the type of the boundary condition for the top surface of the plate. It can be a prescribed *Surface temperature*, a prescribed *Surface heat flux* or convection into a medium at a prescribed *Surrounding temperature*. To set the type of boundary condition, move the cursor over the boundary condition for the top surface. Each time the SPACEBAR is pressed, the type of boundary condition successively changes to *Surface temperature*, *Surface heat flux* and *Surrounding temperature*. In this example the surface temperature is prescribed; therefore, when *Surface temperature* appears on the screen, press the ENTER key in order to set the boundary condition to prescribed *Surface temperature*. Immediately afterwards the cursor moves to the right over the UNIT for the surface temperature. Each time the SPACEBAR is pressed, the UNIT changes, successively, to K, C, and F. In this example the surface temperature is specified in C. Therefore, when C appears on the screen, press the ENTER key in order to set the unit to C. Then the cursor moves to the right, starts blinking and is ready to accept the numerical value of the top surface temperature. In this example it is specified as 830C. Therefore, the value is entered as 830, and then the ENTER key is pressed in order to set the value to 830.

The next item is the *Number of layers*. The program accepts values from 1 to 9 layers. Clearly, 1 corresponds to a single layer plate and 2 to 9 corresponds to a composite plate. In this example the plate has 2 layers. Therefore, the value 2 is entered and the ENTER key is pressed to set the number of layers to 2. Only one integer number is accepted as the number of layers except zero which causes blinking since it has no physical meaning.

The next items are the INPUT data for the *Layer thickness* and the *Thermal conductivity* of layers which are entered by specifying both the UNIT and the MAGNITUDE of these quantities, starting from the first layer. The UNIT for the thickness can be chosen as m, cm, mm, in or ft, and the UNITS for the thermal conductivity can be chosen as W/m·K, W/cm·K, Btu/ft·h·F or kcal/m·h·K. First the cursor is moved over the UNITS for the thickness, and each time the SPACEBAR is pressed, the unit changes successively as stated previously. In this example the thickness is given in meters; therefore, when m appears on the screen, the ENTER key is pressed to set the unit to m. Immediately afterwards the cursor moves to the right, starts blinking and is ready to accept the numerical value of the thickness of the first layer which is 0.2 m. Therefore the value 0.2 is typed and then the ENTER key is pressed. Then the cursor moves immediately over the UNITS for the thermal conductivity. Similarly, the unit is set to W/m·K and then its value 1 for the first layer is entered. By following a similar procedure, the INPUT data regarding the thickness and the thermal conductivity of the second layer are entered by specifying both the units and the magnitudes.

Note that the temperatures appearing on the right are the OUTPUT data for the temperatures of the outer surfaces and the interfaces. Therefore, only the UNITS for these temperatures can be set, because their magnitudes are not known and determined only after the program has been run by pressing the F5 function key. In the present example for prescribed temperatures at the boundary surfaces, the computed boundary temperatures coincide with the specified boundary surface temperatures.

The row next to the last one is for the INPUT data regarding the boundary condition for the bottom surface. This boundary condition can also be a prescribed *Surface temperature*, a prescribed *Surface heat flux* or convection into a medium at a prescribed *Surrounding temperature*. A procedure similar to that used for entering the INPUT data for boundary condition at the top surface is applied for entering the INPUT data for boundary condition at the bottom surface.

For the example considered here the following OUTPUT data are displayed on the screen after the program has been run.

Outer surface temperature at top	=	830C
Interface temperature	=	630C
Outer surface temperature at bottom	=	30C
Heat flow rate	=	1000W

Note that for the case of prescribed surface temperature considered here, the input and the output values of temperatures at the outer boundaries coincide. However, for the case of convection or prescribed heat flux boundary conditions, the temperatures of the outer boundaries are unknown and given by the output data.

Finally, the last row is the OUTPUT data for the heat flow rate through the composite layer. The results for the heat flow rate can be set to any one of the units W, kW, cal/s, kcal/h or Btu/h.

After the program has been run, any of the INPUT data can be changed by moving the cursor over the desired location and entering the new INPUT data. If the new and the old values should be separated by pressing the SPACEBAR. The program is RUN with the new input data by pressing the F5 function key.

Note that, when the INPUT data are changed, the old OUTPUT data remain on the screen until the program is run. Therefore, before printing the screen or copying the output data, it is desirable to press the F5 key to run the program in order to ensure that the OUTPUT data corresponds to the latest INPUT data.

Graphical Support. The results can also be presented in the graphical form if the graphical support is activated by pressing the F8 function key.

The screen displays the coordinate axes "Temperature vs. Distance" over the entire thickness of the composite medium. Computer automatically sets the units, and the ranges of temperature and the distance based on the values specified in the example, i.e., $30 \leq T \leq 830C$ and $0.0 \leq x \leq 0.23$ m. Here x=0.23 m corresponds to the total thickness of the two layers. The user can edit the units as well as the end values of the intervals for temperatures and distances by moving the cursor using F1, F2, F3 or F4 function keys as discussed previously. By the proper choice of the end values for the distances, one can zoom to any one of the layers or any part of them.

To draw the curve press "F8" function key; the following figure will appear on the screen.

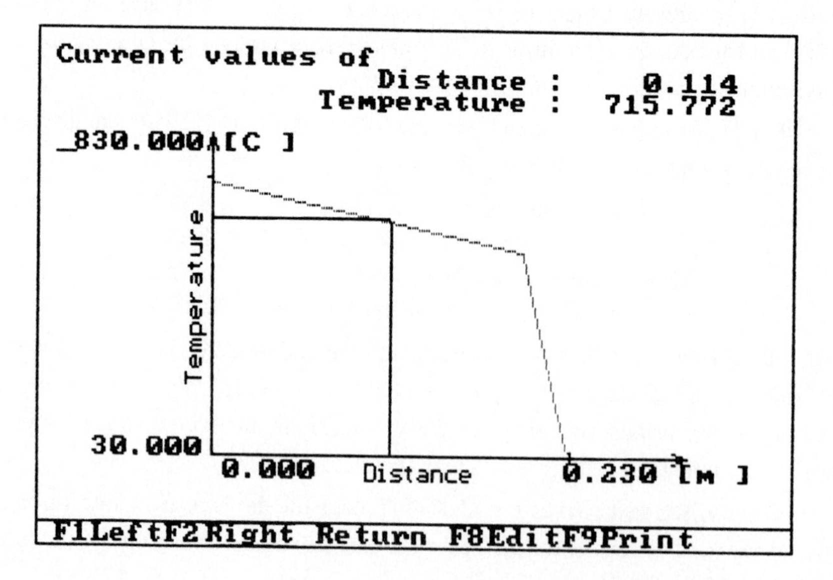

Tabulation of the Results. In the graphical representation of the results, each time the ENTER key is pressed, the *Distance* and *Temperature* corresponding to the cursor position are stored in the memory of the computer. A hard copy of the results stored in the computer memory is obtained by pressing the F9 function key. Table below shows a tabulation of such results.

```
                    COMPOSITE MEDIUM

          Geometry   : Plate

          +-------------------+-------------------+
          |      Distance     |    Temperature    |
          |          [m ]     |            [C ]   |
          +-------------------+-------------------+
          |         0.000     |       830.000     |
          |         0.040     |       789.866     |
          |         0.080     |       749.732     |
          |         0.120     |       709.597     |
          |         0.161     |       669.463     |
          |         0.201     |       630.000     |
          |         0.221     |       215.235     |
          +-------------------+-------------------+
```

Example 2-2. This example is intended to illustrate that in the composite layer problems, the contact resistance between the layers can be treated as a fictitious layer in perfect thermal contact with the neighboring layers.

A composite plane wall consists of two layers with contact resistance (i.e. or contact conductance) between them. The top layer has a thickness L_1=2 cm and a thermal conductivity k_1=0.1 W/m·K. The layer below has a thickness L_2=4 cm and a thermal conductivity k_2=0.05 W/m·K. The top surface is heated by a fluid at temperature T_a=250C with convection with a heat transfer coefficient h_a=15 W/m^2·K. The bottom

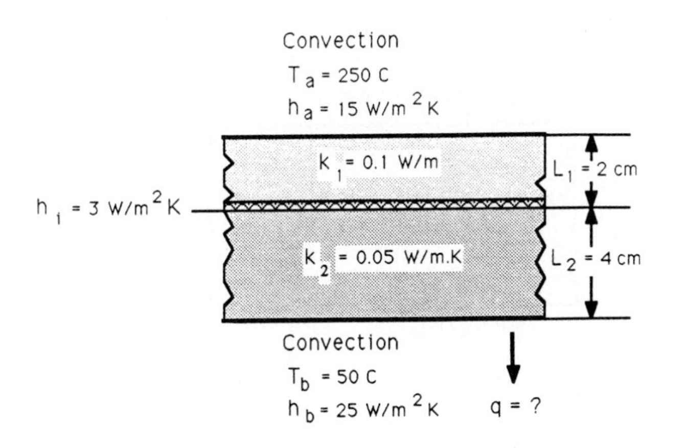

surface is cooled by a cold fluid at temperature T_b=50C with convection with a heat transfer coefficient h_b=25 W/m^2·K. The contact conductance at the interface of the two layers is h_i=3 W/m^2·K. Figure above shows the geometry. Calculate the heat flow rate through the composite layer per meter square, the temperatures of the outer surfaces and of the interfaces.

Solution. This problem differs from that in Example 2-1 because there is convection at the outer boundaries and contact resistance at the interface. The contact resistance can be treated as a fictitious layer which is in perfect thermal contact with the neighboring layers.

The thickness L_i and the thermal conductivity k_i of the fictitious layer is determined from Eq. (2-5), that is

$$\frac{L_i}{k_i} = \frac{1}{h_i}$$

where h_i = 3 W/m^2·K

First the thickness L_i of the fictitious layer is chosen arbitrarily and then the corresponding value of the fictitious thermal conductivity k_i is determined according to the above relation.

Suppose we choose L_i=0.001 m; then k_i is determined from the above expression as

$$k_i = L_i h_i = 0.001 \times 3 = 0.003 \text{ W/m·K.}$$

Thus, the composite medium problem a two layer plate with contact resistance at the interface is now replaced with a three layer slab having all layers in perfect thermal contact as illustrated in the figure below.

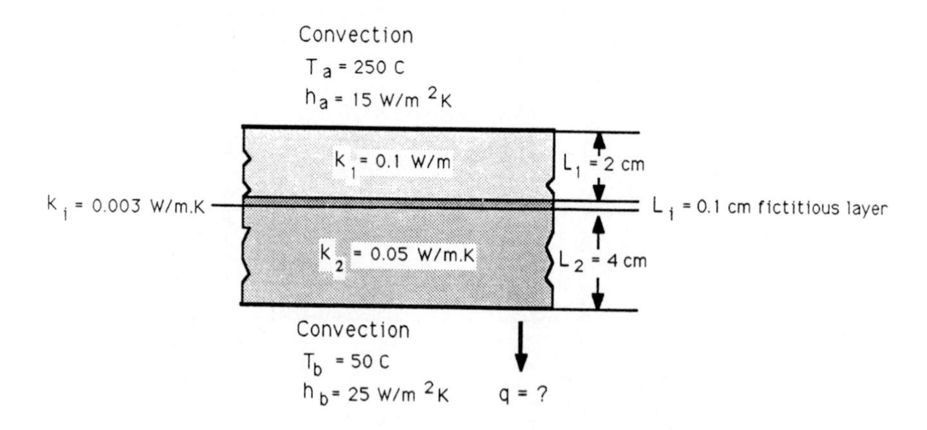

The INPUT and OUTPUT data for this example are shown in the following Table.

```
                      COMPOSITE MEDIUM
                      Geometry : Plate
                          Area : m²                        1.000
      Surrounding temperature : C                        250.000
 Inner heat transfer coefficient : W/m²•K                 15.000

              Number of layers : 3
Layer thickness          Thermal conductivity             Temperatures
cm        2.000     W/m•K                0.100           C       240.741
cm        0.100     W/m•K                0.003           C       212.963
cm        4.000     W/m•K                0.050           C       166.667
                                                         C        55.556



      Surrounding temperature : C                         50.000
 Outer heat transfer coefficient : W/m²•K                 25.000
               Heat flow rate = W                        138.889
```

Note that the order of listing of the INPUT data in this Table from the top to the bottom is similar to the order of listing of the INPUT data in the sketch of the three layer composite wall shown previously. We now describe the main features of entering the INPUT data.

The geometry is selected as *Plate*. The UNIT for the area is chosen as m^2 and magnitude is entered as 1.

The next item is the boundary condition at the top surface of the plate. To set the type of boundary condition, the cursor is moved over the type of boundary condition. Each time the SPACEBAR is pressed, the type of boundary condition successively changes to prescribed *Surface temperature*, prescribed *Surface heat flux* , and convection into a medium at a prescribed *Surrounding temperature.* When *Surrounding temperature* appears on the screen, press the ENTER key to set the boundary condition to convection. Then set the UNIT of temperature to C and its magnitude to 250.

The next item is the entering the UNIT and the magnitude of the *Inner heat transfer coefficient* (i.e., heat transfer coefficient for the top surface). The unit is chosen as $W/m^2{\cdot}K$ and its magnitude is entered as 15.

The problem has three layers, therefore the *Number of layers* is entered as 3. Then the UNITS and magnitudes of the *Layer thickness* and *Thermal conductivity* of each layer are entered successively, starting from the first layer proceeding to the third layer.

The four temperatures following the equal sign "=" are respectively, for the top surface, two interfaces and the bottom surface. These temperatures are unknown, therefore they are the OUTPUT data which will appear on the screen after the program has been run. Therefore, only the UNITS of these four unknown temperatures are selected as C.

The next item is the type of boundary condition for the lower surface. It is set to *Surrounding temperature* which represents convection. The UNIT of the surrounding temperature is set to C and its magnitude to 50.

The *Outer surface heat transfer coefficient* represent the heat transfer coefficient for the lower surface. Its unit is set to $W/m^2 \cdot K$ and magnitude to 25.

For the example considered here, the output data are summarized below.

Heat flow rate	=	138.889W
Top surface temperature	=	240.741C
The first interface temperature	=	212.963C
The second interface temperature	=	166.667C
Bottom surface temperature	=	55.556C.

Finally, the last item in the program, the *Heat flow rate*, appears as the OUTPUT after the program has been run. The heat flow rate can be selected in any one of the following units: W, kW, cal/s, kcal/h or Btu/h. Here it is selected in Watts.

Example 2-3. This example is intended to illustrate that the *COMPOSITE MEDIUM* program can also be used to solve single layer conduction with convection boundary conditions which was not considered in Chapter 1.

A plate of thickness L=5 cm, thermal conductivity k = 2 $W/m \cdot C$ is subjected to convection at both boundary surfaces. The top surface is in contact with a hot fluid at temperature T_a=225C with a heat transfer coefficient h_a=50 $W/m^2 \cdot K$. The bottom surface is in contact with a cold fluid at temperature T_b=25C with a heat transfer coefficient h_b=30 $W/m^2 \cdot K$. Figure below shows the geometry.

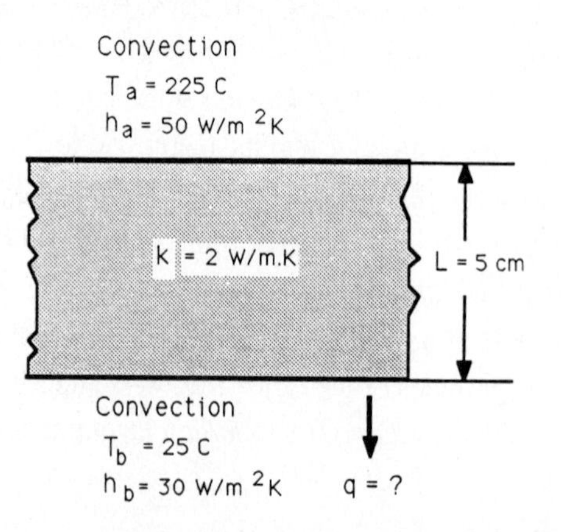

Calculate the heat transfer rate per meter square across the plate and the temperatures of the top and bottom surfaces of the plate.

Solution. Table below shows the computer INPUT and OUTPUT data for this problem. All the input data are entered in a similar manner discussed previously for the composite plate, except the present problem involves only one layer. Since there is convection at both surfaces, the temperatures of the boundary surfaces are not known; they are given as a part of the OUTPUT after the program has been run.

```
                         COMPOSITE MEDIUM
                         Geometry : Plate
                             Area : m²                         1.000
            Surrounding temperature : C                      225.000
      Inner heat transfer coefficient : W/m²•K                50.000

                   Number of layers : 1
      Layer thickness          Thermal conductivity            Temperatures
      cm           5.000    W/m•K               2.000          C      173.936
                                                               C      110.106




            Surrounding temperature : C                        25.000
      Outer heat transfer coefficient : W/m²•K                 30.000
                        Heat flow rate = W                    2553.191
```

As shown in this Table the OUTPUT are

Top surface temperature	=	173.396C
Bottom surface temperature	=	110.106C

Since the area is 1 m^2, the output *Heat flow rate* = 2553.191W is actually the heat flux.

Example 2-4. This example illustrates an application for a two layer hollow cylinder with the layers in perfect thermal contact.

A steam pipe of outside radius 4 cm is covered with a 1 cm thick insulation of thermal conductivity k_1=0.15 W/m·K, which is in turn is covered with a 3 cm thick fiber glass insulation of thermal conductivity k_2=0.05 W/m·K. The layer can be regarded in perfect thermal contact. The steam pipe is kept at 330C and the outside surface of the fiber glass insulation is at 30C. Determine the heat transfer rate per meter length of the pipe and the interface temperature between the asbestos and fiber glass insulations. Figure below shows the geometry.

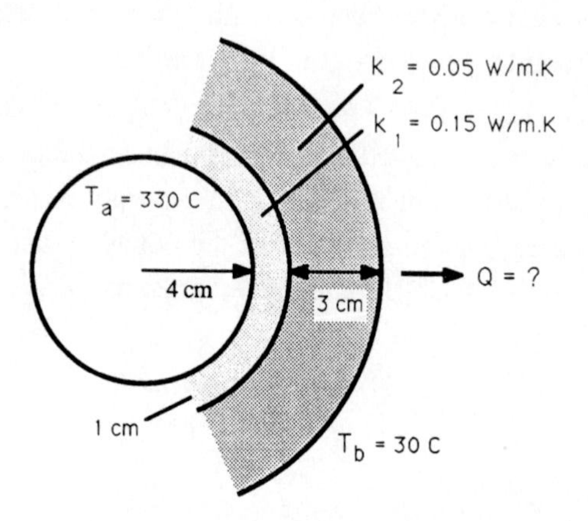

Solution. The following table shows the computer INPUT and OUTPUT data for this problem after the program has been run. The INPUT data for this problem are entered in a similar manner discussed previously for composite slab, but for this case the geometry is cylinder.

```
                    COMPOSITE MEDIUM
                    Geometry : Cylinder
                      Length : m                          1.000
         Surface temperature : C                        330.000

       Inner surface diameter : cm                        8.000
            Number of layers : 2
Layer thickness          Thermal conductivity            Temperatures
cm          1.000     W/m•K               0.150        C      330.000
cm          3.000     W/m•K               0.050        C      289.010
                                                       C       30.000



         Surface temperature : C                         30.000

              Heat flow rate = W                        173.127
```

Example 2-5. This example illustrates the application for the solution of steady-state heat conduction through a two layer cylinder with contact resistance at the interface. The contact resistance is replaced by a fictitious layer in perfect thermal contact with the neighboring layers.

Consider the two layer cylinder problem given in Example 2-4 except there is contact conductance between the layers equal to $h_i=3\ W/m^2 \cdot K$. Figure below shows the geometry.

Solution. This problem differs from the previous example 2-4 in that there is contact resistance at the interface. The contact resistance can be treated as a very thin fictitious cylindrical layer which is in perfect thermal contact with the neighboring layers.

The thickness L_i and the thermal conductivity k_i of such a fictitious layer is determined according to Eq. (2-6), that is

$$\frac{\ln(r_i/r_{i-1})}{k_i} = \frac{1}{h_i}$$

with $h_i = 3\ W/m^2 \cdot K$.

First, the thickness of the fictitious layer $L_i \equiv r_i - r_{i-1}$ is chosen arbitrarily as a quantity much smaller than r_{i-1} and then the corresponding value of the fictitious thermal conductivity k_i is determined according to the above relations. Therefore, we have

$$r_{i-1} = 4 + 1 = 5 \text{ cm.}$$

Suppose we choose $L_i=0.01$ cm, which is much smaller than $r_{i-1}=5$ cm. Then

$$r_i = r_{i-1} + L_i = 5 + 0.01 = 5.01 \text{ cm}$$

and according to the above expression we have

$$\frac{\ln(5.01/5)}{k_i} = \frac{1}{3}$$

from which the thermal conductivity k_i of the fictitious layer is determined as

$$k_i = 0.006\ W/m \cdot K.$$

Therefore, the problem of a two layer cylinder with contact resistance at the interface becomes a problem of 3 layer cylinder with all layers in perfect thermal contact as illustrated in the figure below.

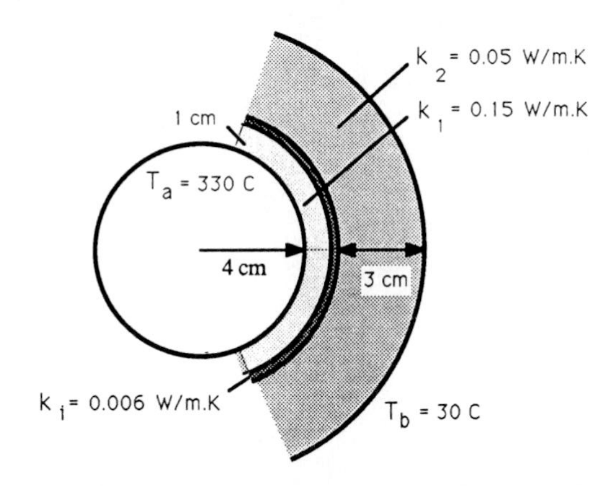

The INPUT and OUTPUT data are shown in the following Table. The OUTPUT data are the heat flow rate, the outer surface and the interface temperatures. In this problem the outer surface temperatures 330C and 30C coincide with the INPUT data because the problem has prescribed outer surface temperatures. The two interface temperatures are 290.17C and 281.258C, and the heat flow rate is 168.214W. Note that this heat flow rate is less than that of Example 2-4 because there is contact resistance between the layers.

```
                         COMPOSITE MEDIUM
                      Geometry : Cylinder
                        Length : m                           1.000
           Surface temperature : C                         330.000

       Inner surface diameter : cm                           8.000
              Number of layers : 3
Layer thickness         Thermal conductivity              Temperatures
cm          1.000     W/m•K                0.150         C       330.000
cm          0.010     W/m•K                0.006         C       290.173
cm          3.000     W/m•K                0.050         C       281.258
                                                         C        30.000




           Surface temperature : C                          30.000

                Heat flow rate = W                         168.214
```

Example 2-6. This example illustrates the application to a two-layer hollow sphere with layers in perfect thermal contact.

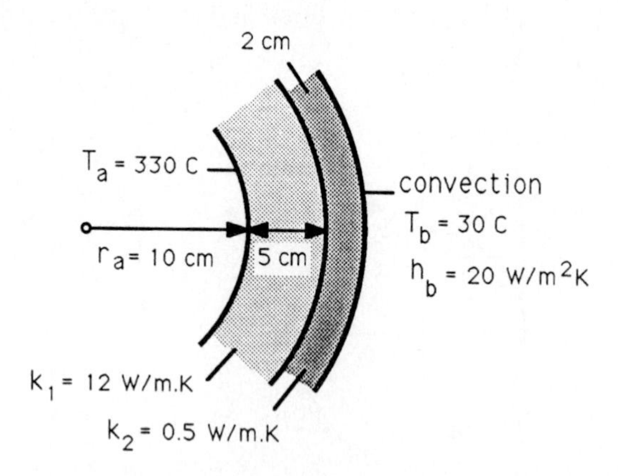

A hollow nickel-steel sphere has an inner radius r_a=10 cm, thickness L_1=5 cm and thermal conductivity k_1=12 W/m·K. It is covered with a L_2=2 cm thick insulation of thermal conductivity k=0.5 W/m·K. The inner surface of the sphere is kept at T_a=230C and the outer surface of the insulation is dissipating heat by convection into a surrounding at T_b=30C with a heat transfer coefficient h_b=20 W/m^2·K. The layers are in perfect thermal contact. Figure above shows the geometry. Calculate the total heat flow rate across the sphere and the interface temperature.

Solution. The computer INPUT and OUTPUT data for this problem after the program has been run is shown in the following Table.

```
                         COMPOSITE MEDIUM
                         Geometry : Sphere

             Surface temperature : C                      330.000

         Inner surface diameter : cm                      20.000
              Number of layers : 2
Layer thickness          Thermal conductivity           Temperatures
cm          5.000    W/m•K              12.000        C      330.000
cm          2.000    W/m•K               0.500        C      306.700
                                                      C      175.122


        Surrounding temperature : C                       30.000
Outer heat transfer coefficient : W/m²•K                  20.000
                Heat flow rate = W                      1054.076
```

This is similar to that for a hollow cylinder, except the geometry is for *Sphere* and there is no need for *Length*. The OUTPUT data include

Inner surface temperature	=	330.0C
Interface temperature	=	306.7C
Outer surface temperature	=	175.122C
Heat flow rate	=	1054.076W

Note that the INPUT and OUTPUT data for the outer surface temperature coincide, because the boundary condition for the outer surface of the sphere is prescribed *Surface temperature.*

Example 2-7. This example illustrates the solution of a composite hollow sphere problem with contact resistance at the interface by representing the contact resistance as a very thin fictitious layer.

Consider the two layer sphere problem of Example 2-6 for the case when there is contact resistance between the layers resulting from an interface contact conductance h_i=50 $W/m^2 \cdot K$. Figure below illustrates the geometry.

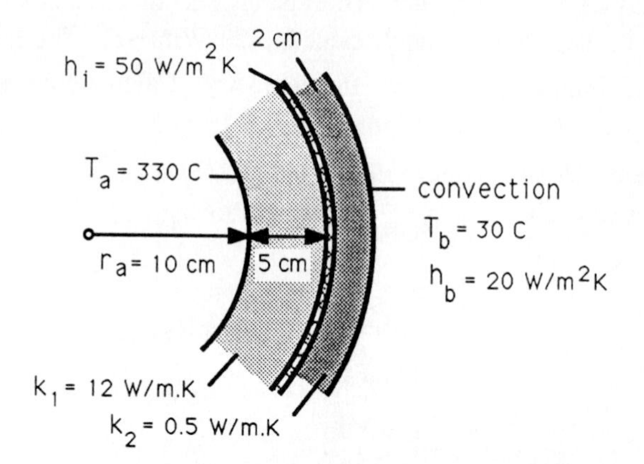

Solution. This problem is exactly the same as that given in Example 2-6, except in the present case there is a contact conductance h_i=50 $W/m^2 \cdot K$ between the layers. Therefore, we need to represent the resulting contact resistance with a very thin fictitious spherical layer which is in perfect thermal contact with the neighboring layer.

The thickness L_i and the thermal conductivity k_i of the fictitious layer is determined according to Eq. (2-7), that is

$$\left(\frac{1}{r_{i-1}} - \frac{1}{r_i}\right)\frac{1}{k_i} = \frac{1}{h_i}$$

with h_i = 50 $W/m^2 \cdot K$.

First, the thickness of the fictitious layer $L_i = r_i - r_{i-1}$ is chosen arbitrarily as a quantity much smaller than r_{i-1} and then the corresponding value of the fictitious thermal conductivity k_i is determined from the above expression. Therefore, we have

$$r_{i-1} = 10 + 5 = 15 \text{ cm.}$$

Suppose we choose L_i=0.1 cm, which is much smaller than r_i=15 cm. Then

$$r_i = r_{i-1} + L_i = 15 + 0.1 = 15.1 \text{ cm.}$$

and the fictitious thermal conductivity k_i is determined according to the above expression as

$$\left(\frac{1}{15} - \frac{1}{15.1}\right)\frac{1}{k_i} = \frac{1}{50}$$

or $\quad k_i = 0.022$ W/m·K.

Having established the thickness L_i and thermal conductivity k_i of the fictitious layer, the above problem of two layer sphere with contact resistance at the interface is replaced by a 3 layer sphere problem with all layers in perfect thermal contact as illustrated in the following figure.

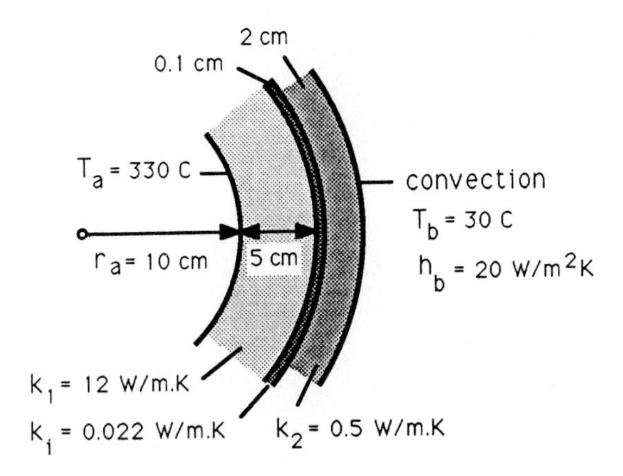

The computer INPUT and OUTPUT for this problem after the program has been run is given in the following Table.

```
                   COMPOSITE MEDIUM
                   Geometry : Sphere

         Surface temperature : C                           330.000

      Inner surface diameter : cm                           20.000
            Number of layers : 3
Layer thickness         Thermal conductivity              Temperatures
cm          5.000    W/m•K             12.000            C       330.000
cm          0.100    W/m•K              0.022            C       314.968
cm          2.000    W/m•K              0.500            C       206.367
                                                         C       122.534


     Surrounding temperature : C                            30.000
Outer heat transfer coefficient : W/m²•K                    20.000
              Heat flow rate = W                           680.041
```

A comparison of this result with that for the case of two layer sphere with layers in perfect contact reveals that the presence of contact conductance reduced the total heat flow rate from 1054.076W to 680.041W. Also, the center surface temperature is reduced from 175.22C to 122.534C.

Chapter

THREE

LUMPED SYSTEM ANALYSIS

There are many practical applications in which the variation of temperature within the solid with position is negligible, hence the temperature is considered to vary only with time. The analysis of unsteady heat flow under such an assumption, generally referred to as *lumped system analysis*, greatly simplifies the problem of determining the solid temperature as a function of time. The range of applicability of such a simple analysis is very limited; but when applicable, it provides a very quick answer to the temperature variation of the solid with time.

1-3 ANALYSIS

Consider a solid of arbitrary shape of volume V, total surface area A, thermal conductivity k, density ρ, specific heat C_p, initially at a uniform temperature T_0. Suddenly, at time t=0, the solid is exposed to convective heat transfer with an ambient at a constant temperature T_∞ with a heat transfer coefficient h.

The temperature distribution during transients within the solid at any instant is uniform, with an error less than about 5 percent if the following criteria is satisfied:

$$Bi = \frac{hL_s}{k} < 0.1$$

where Bi is called the Biot number and the *characteristic length* L_s of the solid is defined as

$$L_s = \frac{\text{Volume}}{\text{Surface area}} = \frac{V}{A}$$

Based on the lumped system analysis, the variation of temperature of the solid with time is given by

$$T(t) = T_\infty + (T_o - T_\infty)\exp\left(-\frac{Ah}{\rho C_p V}\,t\right)$$

where

T_∞ = ambient temperature

T_o = initial temperature
A = surface area
h = heat transfer coefficient
t = time
ρ = density
C_p = specific heat
V = volume

3-2 COMPUTER SOLUTIONS

We illustrate the application of the lumped system analysis with representative examples.

The input data are entered as discussed in the previous chapters. In the program *LUMPED SYSTEM ANALYSIS*, either the *time* or the *temperature* can be the output if followed by the equal sign.

To change the column sign into the equal sign, move the cursor over the column sign and press the SPACEBAR.

Example 3-1. An aluminum plate of volume V=300 cm^3 and surface area A=200 cm^2 is to be cooled from 225C to 50C by convection into an ambient at a constant temperature T_∞=25C. If the heat transfer coefficient is h=320 $W/m^2{\cdot}K$, how long it will take for the plate to cool to 50C. The properties can be taken as: ρ=2790 kg/m^3, C_p=880 J/kg·K and k=160 W/mK.

Solution. Table below shows the computer INPUT and OUTPUT data for this problem. In the problem considered here, the time is unknown, hence is the unknown quantity followed by the equal sign. The answer for the problem is t=3.989 min.

LUMPED SYSTEM ANALYSIS

Volume	:	cm^3	300.000
Surface area	:	cm²	200.000
Density	:	kg/m^3	2790.000
Specific heat	:	J/kg•K	880.000
Thermal conductivity	:	W/m•K	160.000
Heat transfer coefficient	:	W/m²•K	320.000
Initial temperature	:	C	225.000
Ambient temperature	:	C	25.000
Time	=	min	3.989
Temperature	:	C	50.000

Graphical Support To present the results in the graphical form, press the F8 function key in order to activate the graphical support. The screen displays the coordinate axes *Temperature* vs. *Time* and automatically sets the units and the ranges for the temperature and time based on the values specified above, i.e., $0 \leq t \leq 3.989$ min and $25 \leq T \leq 225$C.

The units as well as the end values of these intervals can be edited by moving the cursor over the quantity to be edited by using the F1, F2, F3 and F4 function keys.

In this example the values automatically set by the computer are used except for the end of the time interval which is altered from the 3.989 min. to 4 min.

Now the computer is ready to calculate the temperature and plot the results. Press the F8 function key to start calculations. The following figure will appear on the screen.

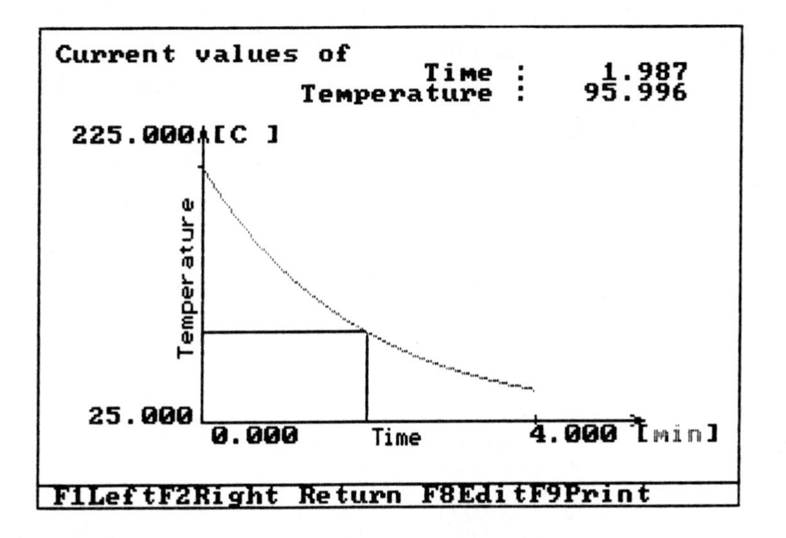

Tabulation of the Results. The exact values of temperature and time corresponding to the cursor location on the graph are displayed at the upper right corner of the graph. In the above figure the cursor location corresponds to a temperature of 96.206C at time 1.981 minute after the start of cooling.

In the graphical form, each time the ENTER key is pressed, the *Temperature* and *Time* corresponding to the cursor location are stored in the computer memory. Up to 150 such points can be stored quantities can be obtained by pressing the F9 function key. The following Table shows such a tabulation.

LUMPED SYSTEM ANALYSIS

Time [min]	Temperature [C]
0.000	225.000
0.510	178.300
1.020	142.505
1.503	116.337
2.013	95.010
2.497	79.419
3.007	66.712
3.490	57.423
4.000	49.852

Example 3-2. We now consider Example 3-1 for the case when the temperature of the body is to be determined at a specified time. That is, we have the same problem as specified in Example 3-1, but ask for the temperature of the solid at the time t = 5 minutes after the start of the cooling.

Solution. Table below shows the computer INPUT and OUTPUT data for this specific problem. The final temperature being unknown the equal sign is placed after the *Temperature* by moving the cursor over the column sign in front of temperature and pressing the SPACEBAR. The answer for the problem is T = 39.755C.

LUMPED SYSTEM ANALYSIS

Volume :	cm^3	300.000
Surface area :	cm²	200.000
Density :	kg/m^3	2790.000
Specific heat :	J/kg•K	880.000
Thermal conductivity :	W/m•K	160.000
Heat transfer coefficient :	W/m²•K	320.000
Initial temperature :	C	225.000
Ambient temperature :	C	25.000
Time :	min	5.000
Temperature =	C	39.755

Chapter

FOUR

TRANSIENT CONDUCTION

The determination of the temperature distribution within the solid during temperature transients is a more complicated matter than the solution of the steady-state problems or the lumped systems because temperature varies with both position and time. For example, if the surface temperature of a solid body is suddenly altered, the temperature within the body begins to change over time; it will take sometime before the steady-state temperature distribution is again established.

One-dimensional transient heat conduction problems for solids having simple shapes such as a plate, solid cylinder or solid sphere are readily solved with analytical approaches.

In this chapter we present analytic expressions for one-dimensional transient temperature distribution in solids having geometries such as a semiinfinite medium confined to the region $x \geq 0$, a flat plate of thickness 2L, along solid cylinder of radius L and a solid sphere of radius L as illustrated in the figure below. It is assumed that there is no energy generation in the medium, thermal properties are constant, and the solid is

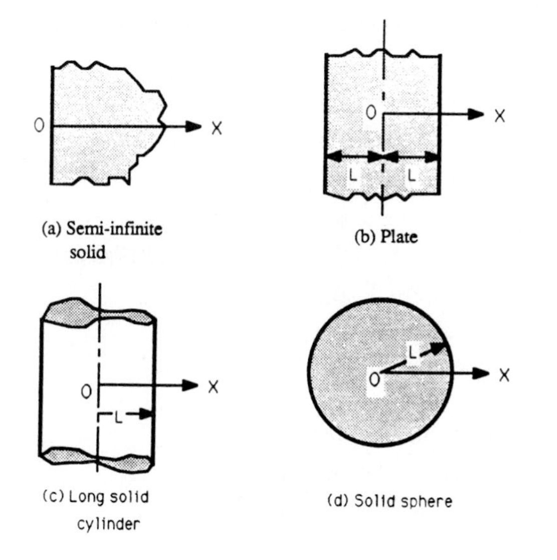

Figure 4-1 Geometry and Coordinates

initially at a uniform temperature T_o. For each of the four geometries shown in Fig. 4-1, the following three different boundary conditions are considered:

(i) Constant temperature T_w applied to the boundary surface,

(ii) Constant heat flux q applied to the boundary surface, and

(iii) Convection into a surrounding fluid at a specified constant temperature T_f with a heat transfer coefficient h.

4.1 SEMIINFINITE SOLID

Analytical expressions are available for the one-dimensional transient temperature distribution in various textbooks such as by Carslaw and Jaeger (1959), Ozisik (1980) and Mikhailov and Ozisik (1984). We summarize below these expressions for each of the three different boundary conditions described above.

1. Constant surface temperature T_w applied at the surface x=0:

$$T(x,t) = T_w + (T_o - T_w)\,\mathrm{erf}\left(\frac{x}{2\sqrt{\alpha t}}\right) \tag{4-1}$$

2. Constant surface heat flux q applied at the surface x=0:

$$T(x,t) = T_o + \frac{2q}{k}\sqrt{\frac{\alpha t}{\pi}}\,\exp\left(-\frac{x^2}{4\alpha t}\right) - \frac{qx}{k}\,\mathrm{erfc}\left(\frac{x}{2\sqrt{\alpha t}}\right) \tag{4-2}$$

3. Convection at the surface x=0 into a surrounding at constant temperature T_f with a heat transfer coefficient h:

$$T(x,t) = T_o + (T_f - T_o)\left\{\mathrm{erfc}\left(\frac{x}{2\sqrt{\alpha t}}\right) - \exp\left(\frac{hx}{k} + \frac{h^2\alpha t}{k^2}\right)\mathrm{erfc}\left(\frac{x}{2\sqrt{\alpha t}} + \frac{h\sqrt{\alpha t}}{k}\right)\right\} \tag{4-3}$$

where erf(x) and erfc(x) are, respectively, the error function and the complimentary error function of argument x, and various other quantities are defined as

h	=	heat transfer coefficient
k	=	thermal conductivity
q	=	applied surface heat flux
t	=	time
T_o	=	initial temperature
T_f	=	surrounding temperature
T_w	=	applied surface temperature

4.2 PLATE, CYLINDER AND SPHERE

Analytical solutions are given for one-dimensional transient temperature distribution in a plate of thickness 2L, in a solid cylinder or sphere of radius L in various texts such as by Carslaw and Jaeger (1959), Ozisik (1980) and Mikhailov and Ozisik (1984). These expressions are in the form of infinite series. In the case of a plate of thickness 2L, it is assumed that the problem has a geometric and thermal symmetry about the central plane.

The analytic solution given below are applicable to plate, solid cylinder and solid sphere if the indices "m" appearing in these equations are chosen as

$$m = \begin{cases} \frac{1}{2} & \text{for plate} \\ 0 & \text{for cylinder} \\ -\frac{1}{2} & \text{for sphere} \end{cases}$$

The resulting expressions for each of the three different boundary conditions are:

1. Constant surface temperature T_w :

$$T(X,\tau) = T_w + (T_o - T_w)\, 2 \sum_{i=1}^{\infty} X^m \frac{J_{-m}(\mu_i X)}{\mu_i J_{1-m}(\mu_i)} e^{-\mu_i^2 \tau} \tag{4-4}$$

where the eigenvalues μ_i are the roots of the transcendental equation

$$J_{-m}(\mu) = 0 \tag{4-5}$$

in addition

T_o = initial temperature

T_w = applied surface temperature

$X = \frac{L-d}{L}$, with L is half-thickness of plate, radius of cylinder and sphere, d is distance from the surface

$\tau = \frac{\alpha t}{L^2}$, dimensionless time with α being thermal diffusivity and t time.

2. Constant surface heat flux, q :

$$T(X,\tau) = T_o + \frac{qL}{k}\left[2(1-m)\tau + \frac{1}{2}\left(X^2 - \frac{1-m}{2-m}\right) - 2\sum_{i=1}^{\infty} X^m \frac{J_{-m}(\mu_i X)}{\mu_i^2 J_{-m}(\mu_i)} e^{-\mu_i^2 \tau}\right] \tag{4-6}$$

where the eigenvalues μ_i are the roots of the transcendental equation

$$J_{1-m}(\mu) = 0 \tag{4-7}$$

and q = applied surface heat flux

k = thermal conductivity.

3. Convection at the boundary surface.

$$T(X,\tau) = T_o + (T_w - T_o)$$

$$\cdot \left(1 - 2\sum_{i=1}^{\infty} \frac{1}{\mu_i}\left[1 + \frac{2m}{Bi} + \left(\frac{\mu_i}{Bi}\right)^2\right]^{-1} X^m \frac{J_{-m}(\mu_i X)}{J_{1-m}(\mu_i)} e^{-\mu_i^2 \tau} \right) \tag{4-8}$$

where the eigenvalues μ_i are the roots of the transcendental equation

$$\frac{J_{-m}(\mu)}{J_{1-m}(\mu)} = \frac{\mu}{Bi} \tag{4-9}$$

and $Bi = \frac{hL}{k}$ is the Biot number, with h being the heat transfer coefficient.

The functions $X^m J_{-m}(X)$ and $X^m J_{1-m}(X)$ appearing in the above equations for plate ($m = \frac{1}{2}$), cylinder ($m = 0$) and sphere ($m = -\frac{1}{2}$) are given in the following table

Table 4-1

m	$X^m J_{-m}(X)$	$X^m J_{1-m}(X)$
$\frac{1}{2}$	$\sqrt{\frac{2}{\pi}}\cos X$	$\sqrt{\frac{2}{\pi}}\sin X$
0	$J_0(X)$	$J_1(X)$
$-\frac{1}{2}$	$\sqrt{\frac{2}{\pi}}\, j_0(X)$	$\sqrt{\frac{2}{\pi}}\, j_1(X)$

where the spherical Bessel functions of the first kind, $j_0(X)$ and $j_1(X)$ are

$$j_0(X) = \frac{\sin X}{X} \quad \text{and} \quad j_1(X) = \frac{\sin X - X \cos X}{X^2}$$

4-3 COMPUTER SOLUTIONS

The analytic expressions given previously are used in the program *TRANSIENT CONDUCTION* to calculate one-dimensional transient temperature distribution in geometries such as a *Semiinfinite medium*, *Plate*, *Cylinder* and *Sphere*, initially at a uniform temperature T_0, and for times $t>0$, the boundary surface is subjected to any one of the three different boundary conditions discussed previously. In the case of plate, cylinder, and sphere the solutions are in the form of infinite series which requires the evaluation of large number of terms in the series for times $t \to 0$. In this software, only the first ten terms of the summation are computed, as a result the solutions for t=0 or extremely small times differ from the initial temperatures. The computation of transient temperatures for plate, solid cylinder and sphere also requires the solution of transcendental equations in order to calculate the eigenvalues for the problem. The roots of the transcendental equations are determined by using approximations in terms of Chebyshev polynomials discussed by Mikhailov and Shishedjiev (1975).

Only 10 terms are used in the summation hence the solution is not applicable for t=0.

The graphical support is also provided in order to plot the temperature at any point in the medium as a function of time over any specified time interval.

We now illustrate the application of the program *TRANSIENT CONDUCTION* for determining the temperature of the solid at any specified location for any given time. Conversely, the solution of the problem involving the determination of the time required for any point in the solid to attain a specified temperature requires iteration. The use of manual iteration and the graphical support to perform the iteration are also discussed.

Example 4-1. A very thick concrete wall can be approximated as a semi-infinite medium having the boundary surface at $x=0$. Initially the wall is at a uniform temperature $T_0=125C$. At time t=0, the temperature of the surface at $x=0$ is suddenly lowered $T_w=25C$ and maintained at that temperature for $t \geq 0$. The wall has a thermal diffusivity $\alpha=7\times10^{-7}\ m^2/s$. (i) Calculate the temperature of the solid at the location $x=5$ cm from the surface at time $t=30$ min after the surface is subjected to cooling, (ii) Plot the temperature at this location as a function of time over the time interval from t=0 to $t=30$ min, and (iii) present the results in a tabulated form. Figure below shows the geometry and the coordinate.

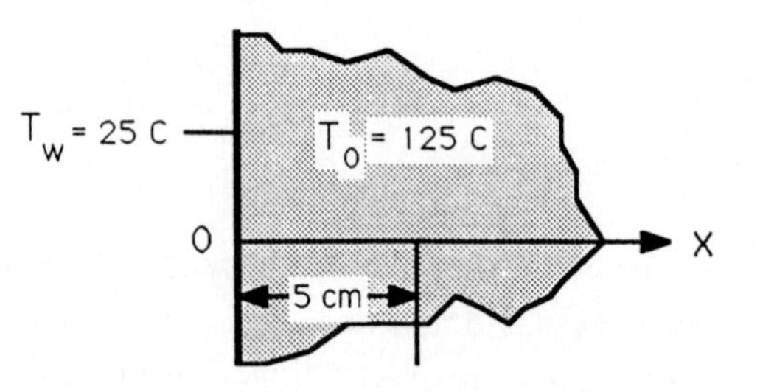

Solution. Table below shows the computer INPUT and OUTPUT data for this problem after the program has been run. The main features of entering the INPUT data, the use of the GRAPHICAL support and obtaining a hard copy of tabulated data are now described.

TRANSIENT CONDUCTION

Geometry	: Semiinfinite medium	
Initial temperature	: C	125.000
Surface temperature	: C	25.000
Thermal diffusivity	: m²/s	0.00000070
Distance from the surface	: cm	5.000
Time	: min	30.000
Temperature	= C	93.076

Entering the INPUT data. The first INPUT data in Table above is the *Geometry*. To set the type of geometry, move the cursor next to the *Geometry* and press the SPACEBAR. Each time the SPACEBAR is pressed the type of geometry changes, successively, to *Semiinfinite medium, Plate, Cylinder* and *Sphere*. Press the ENTER key when the desired geometry is displayed. For this problem it is set to *Semiinfinite medium.*

The next item is the *Initial temperature*. First the unit of temperature is selected and then its magnitude is entered. Move the cursor over the UNIT, press the SPACEBAR. Each time the SPACEBAR is pressed the unit changes, successively, to K, C and F. Press the ENTER key when the desired unit appears. In this example Celsius is chosen. Then the cursor moves to the right, starts blinking and is ready to accept the value of temperature. Type the value (i.e., 125), press the SPACEBAR to ensure that the newly entered value is separated from the old one and then press the ENTER key.

The next item is the selection of the type of boundary condition. Move the cursor over the boundary condition and press the SPACEBAR. Each time the SPACEBAR is pressed, the type of boundary condition changes, successively, to prescribed *Surface temperature*, prescribed *Surface heat flux* and convection into an ambient at a prescribed *Surrounding temperature*. When the desired boundary condition appears, press the ENTER key. In this Example, the boundary condition is a prescribed *Surface temperature*. Then select the unit for the temperature and then type its value as discussed previously. In this example it is 25C.

The next item is the selection of the units of the thermal diffusivity. Move the cursor over *Thermal diffusivity* and press the SPACEBAR. Each time the SPACEBAR is pressed the unit changes, successively, to m^2/s, cm^2/s and ft^2/s. Press the ENTER key when the desired unit appears. Then the cursor moves to the right, starts blinking, and is ready to accept the value of the thermal diffusivity. Enter the value either in decimals (i.e., 0.0000007) or in exponential (i.e., 7e-7), press the SPACEBAR to separate the new value from the old one and then press the ENTER key.

The next INPUT is the *Distance from the surface* of the point at which temperature is to be determined. Move the cursor over the UNIT and press the SPACEBAR. Each time the SPACEBAR is pressed the unit changes, successively, to m, cm, mm, in and ft when the desired unit appears press the ENTER key. Then type the value of the distance, press the SPACEBAR and then the ENTER key. In this Example the distance is 5 cm.

The next INPUT data is the *Time* at which the temperature at the selected location is to be calculated. The *Time* can be selected in any one of the units s, min and h. In this Example it is 30 min.

The program is now ready to be run. Press the F5 function key to RUN the program. The answer to the problem is 93.076C.

Finally, the last item *Temperature* is the computer OUTPUT, therefore it is followed by an equal sign "=". Move the cursor over the unit for temperature and press the SPACEBAR. Each time the SPACEBAR is pressed the output becomes, successively, in the units of K, C and F. In this Example C is chosen for the unit of the temperature output.

Note that the unit of anyone of the INPUT and OUTPUT data can also be changed after the program has been run. The value is automatically adjusted to the one corresponding to the new unit. However, the program does not accept zero as an input for time.

Graphical Support. To present the results in the graphical form, press the F8 function key in order to activate the graphical support. The screen displays the coordinate axes *Temperature* vs. *Time*, automatically sets the units and the ranges for the temperature and the distance based on the values specified in the example, i.e., $25 \leq T \leq 125$C and $0 \leq t \leq 30$ min. The units as well as the end values of these intervals can be edited by moving the cursor over the quantities to be modified by using the F1, F2, F3 and F4 function keys and entering the desired values as described previously.

In this example the range of time is specified as $0 \leq t \leq 30$ min and the initial range of temperature as $25 \leq T \leq 125$C. The computer automatically selects these ranges for the graphical representation. To plot the results, press F8 key. Suppose we wish to examine the temperature distribution over a much narrower range, say, $90 < T < 125$C. In such a case, one needs to change from the Draw to the Edit mode by pressing again the F8 key. This key serves for two purposes, namely, to draw the curve and to change to the edit mode. At any time, the current function of the F8 key is indicated on the screen. Therefore, move the cursor over the lower value 90C, change it to 25C and press the F8 function key to replot the results. The following figure is obtained

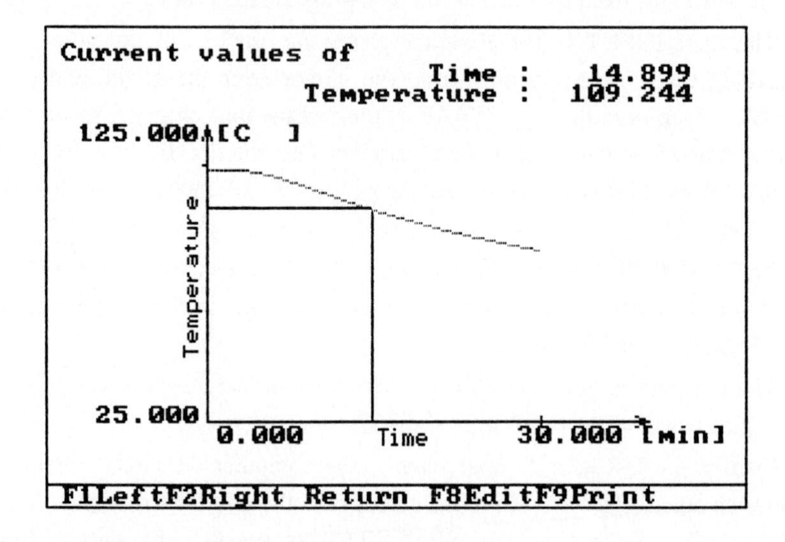

Tabulation of the Results. In the above figure the cursor location corresponds to T=109.244C at the time t=14.899 min after the start of cooling. If the cursor is moved along the graph, the current values of temperature and time corresponding to the cursor location automatically replace the old ones. Each time the ENTER key is pressed, the *Temperature* and *Time* corresponding to the cursor location are stored in the computer memory. Up to 150 such values can be stored since the computer divides the time interval into 150 equal parts and calculates the temperature at each of these times. A hard copy of these results can be obtained by pressing the F9 function key. Table below shows a tabulation of such results

TRANSIENT CONDUCTION

Geometry : Semiinfinite medium

Time [min]	Temperature [C]
0.000	125.000
5.034	123.497
10.067	116.446
15.101	108.965
20.134	102.594
25.168	97.316
30.000	93.076

Example 4-2. A very thick concrete wall can be approximated as a semiinfinite medium having its boundary surface at X=0. Initially the wall at a uniform temperature T_o = 125C. At time t=0, the boundary surface at X=0 is subjected convection with a surrounding fluid at temperatureT_f = 25C with a heat transfer coefficient h = 8 W/m^2·K. The concrete has a thermal diffusivity $\alpha = 7 \times 10^{-7}$ m^2/s and a thermal conductivity k=1 W/m·K. Figure below shows the geometry and coordinates. Calculate the temperature at a location X=5 cm from the surface at time t=30 min after the surface subjected to cooling.

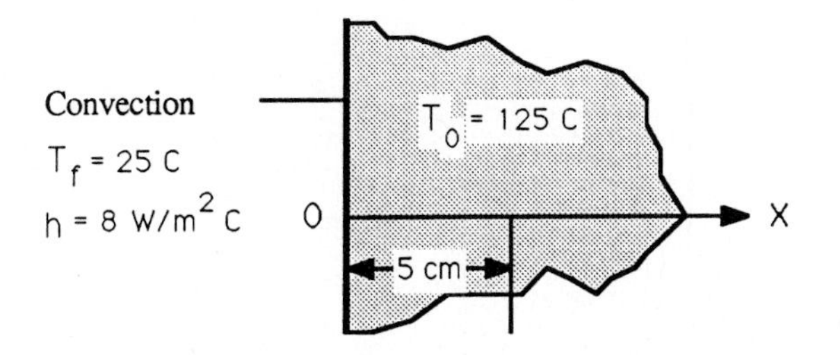

Solution. Table below shows the computer INPUT and OUTPUT data for this problem after the program has been run.

```
                    TRANSIENT  CONDUCTION

                   Geometry : Semiinfinite medium
        Initial temperature : C                       125.000

    Surrounding temperature : C                        25.000
  Heat transfer coefficient : W/m²•K                    8.000

       Thermal conductivity : W/m•K                     1.000
        Thermal diffusivity : m²/s                 0.00000070
  Distance from the surface : cm                        5.000
                       Time : min                      30.000

                Temperature = C                       119.310
```

Various INPUT data are entered in a similar manner described in the previous example, except the boundary condition at the surface X=0 is now convection. Therefore, when the cursor is over the boundary condition press the SPACEBAR successively until *Surrounding Temperature* is displayed. Then press the ENTER key. Enter the unit and the value of the surrounding temperature which is 25C.

The following three items are the *Heat transfer coefficient, Thermal conductivity* and *Thermal diffusivity.* Enter first the unit and then the value of each of these quantities. The last item is the computer OUTPUT for temperature which provides the answer for the problem (i.e., 119.31C).

Graphical Support. Figure below shows a graphical representation of temperature over the time interval from 0 to 2 hr. The temperature range selected is $90 \leq T \leq 125$C.

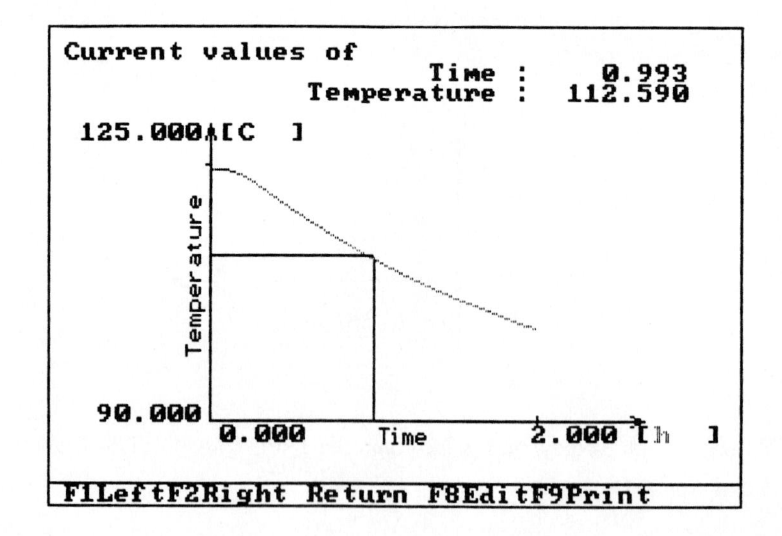

A comparison of the present result with that of Example 4-1 shows that the cooling rate with convection boundary condition is slower than that with the prescribed temperature boundary condition.

Example 4-3. A 3cm thick marble plate of thermal diffusivity $\alpha=1.3\times10^{-6}$ m^2/s and thermal conductivity k=3 W/m·K is initially at a uniform temperature T_0=130C. At time t=0, both boundary surfaces are suddenly lowered to 30C. Calculate the center-plane temperature at time 2 minutes after the lowering of boundary surface temperatures. The following figure shows the geometry and coordinate.

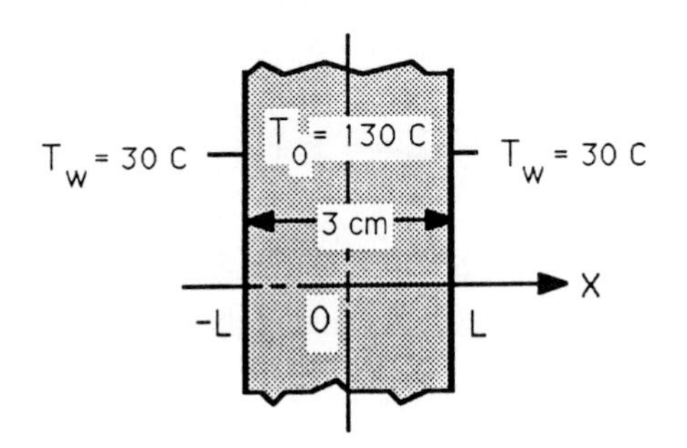

Solution. Table below shows the computer INPUT and OUTPUT data for this example after the program has been run.

```
                 TRANSIENT   CONDUCTION

            Geometry : Plate
 Initial temperature : C                    130.000

 Surface temperature : C                     30.000

      Half-thickness : cm                     1.500

  Thermal diffusivity : m²/s             0.00000130
Distance from the surface : cm                1.500
                  Time : min                  2.000

          Temperature = C                    53.012
```

All the INPUT data are entered in a similar manner discussed previously. The last item is the OUTPUT which is the solution to the problem. That is, the centerplane temperature at a time 2 min after the start of the transients is 53.012C.

Example 4-4. A very thick copper plate of thermal diffusivity $\alpha=1.1\times10^{-5}$ m^2/s and thermal conductivity k=386 W/m·K is initially at a uniform temperature

T_o=25C. Suddenly, one of its surfaces is exposed to a constant heat flux $q_o=2\times10^5$ W/m^2. Assuming that the plate can be regarded as a semiinfinite medium, calculate the temperature at a location x=2 cm from the surface at a time t=5 mins after the application of the surface heating. Figure below shows the geometry and coordinate.

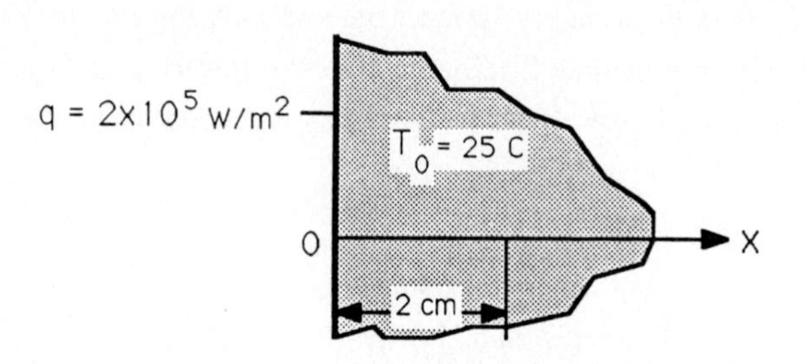

Solution. The following Table shows the computer INPUT and OUTPUT data for the solution of this problem after the program has been run.

```
TRANSIENT  CONDUCTION

              Geometry : Semiinfinite medium
   Initial temperature : C                    25.000

     Surface heat flux : W/m²             200000.000


  Thermal conductivity : W/m•K               386.000
   Thermal diffusivity : m²/s             0.00001100
Distance from the surface : cm                 2.000
                  Time : min                   5.000

           Temperature = C                    49.236
```

All the input data are entered in a similar manner as discussed previously, except the boundary condition for this problem is a prescribed *Surface heat flux.* The last item in this Table is the computer OUTPUT which gives the temperature at the location x=2 cm from the surface as 49.236C at a time t=5 minutes after the application of the surface heat flux.

Example 4-5. An aluminum plate of thickness L=4 cm, thermal diffusivity $\alpha=8\times10^{-6}$ m^2/s is initially at a uniform temperature T_o=20C. At time t=0, suddenly one of its surfaces is raised to T_w=220C while the other surface is kept insulated. Calculate the temperature of the insulated surface at a time t=2 min after the start of the temperature transients. Figure below shows the geometry and coordinate.

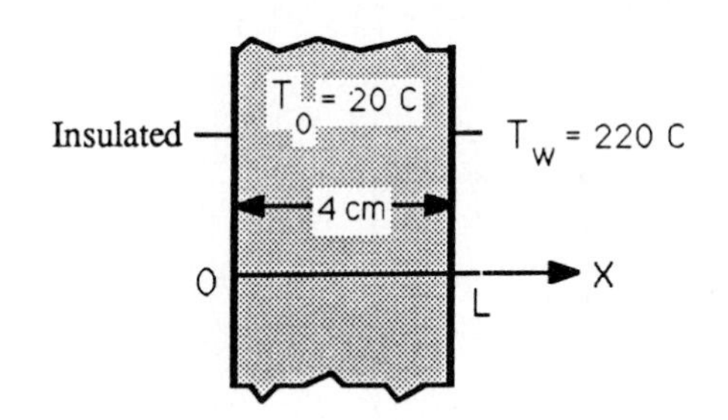

Solution. This problem has thermal and geometric symmetry about the insulated boundary surface at x=0. Therefore it is equivalent to the transient heat conduction problem of a plate of thickness 2L. Therefore, the program for a plate, initially at a uniform temperature and for times t>0 both boundary surfaces are subjected to a prescribed surface temperature is applicable for its solution.

Table below shows the computer INPUT and OUTPUT data for this problem after the program has been run.

```
                    TRANSIENT  CONDUCTION

                  Geometry : Plate
       Initial temperature : C                     20.000

       Surface temperature : C                    220.000

            Half-thickness : cm                     4.000

      Thermal diffusivity : m²/s              0.00000800
 Distance from the surface : cm                     4.000
                      Time : min                    2.000

               Temperature = C                    162.058
```

The INPUT data are entered as discussed previously. For this problem the half-thickness of the plate is 4 cm and the location of the insulated surface is 4 cm from the boundary surface. Finally, the last item in this Table is the computer OUTPUT which gives the solution for the problem as 162.058C at time 2 min after the start of heating.

Example 4-6. A solid iron rod of diameter D=6 cm, thermal diffusivity $\alpha=2\times10^{-5}$ m^2/s and thermal conductivity k=60 W/m·K is initially at a uniform temperature T_0=800C. Suddenly the rod is immersed into an oil bath at temperature T_∞=50C. The heat transfer coefficient between the oil and the rod surface is h=400

W/m^2·K. Determine the centerline temperature at a time 5 mins after the immersion of the rod into the fluid. Figure below shows the geometry.

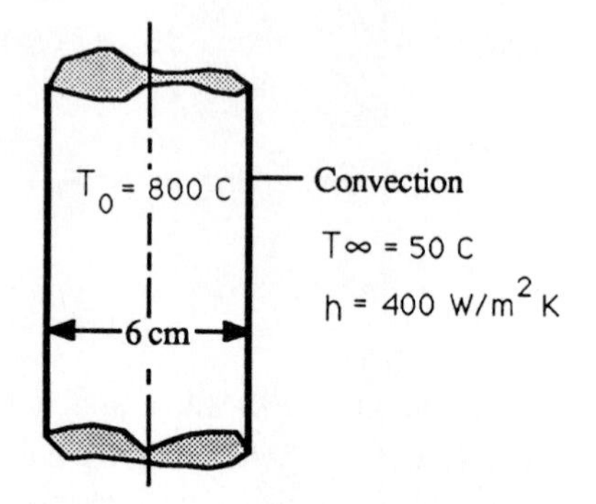

Solution. The following Table shows the computer INPUT and OUTPUT data for this problem.

TRANSIENT CONDUCTION

Geometry	:	Cylinder	
Initial temperature	:	C	800.000
Surrounding temperature	:	C	50.000
Heat transfer coefficient	:	W/m²·K	400.000
Radius	:	cm	3.000
Thermal conductivity	:	W/m·K	60.000
Thermal diffusivity	:	m²/s	0.00002000
Distance from the surface	:	cm	3.000
Time	:	min	5.000
Temperature	=	C	112.145

The last item is the computer OUTPUT which shows that the centerline temperature at time 5 minutes after the start of cooling is 112.145C.

Example 4-7. A solid iron rod of diameter D=6 cm, thermal diffusivity $\alpha=2\times10^{-5}$ m^2/s and thermal conductivity k=60 W/m·K is initially at a uniform temperature T_o = 800K. Suddenly the rod is immersed into an oil bath at temperature T_∞=50C. The heat transfer coefficient between the oil and the rod surface is h=400 W/m^2·K. How long it will take the centerline temperature to reach 120C?

Solution. This problem is similar to that considered in Example 4-6, except time is required for the centerline temperature to reach 120C. Therefore, the solution cannot be obtained directly; iteration is needed. Here we consider two different approaches for the solution: (i) a manual iteration, and (ii) using a graphical support for iteration.

Solution by Manual Iteration. The problem is solved as in Example 4-6 for several different times until the centerline temperature determined in this manner is sufficiently close to 120C. For example,

Time, secs	Centerline Temp., C
285.0	120.552
286.0	119.958
285.6	120.195
285.8	120.077

Iteration by Graphical Support. First the problem is solved as in Example 4-6 to determine the centerline temperature at a specified time and the results are presented in the graphical form by plotting the centerline temperature against time over the time interval from 0 to 300 sec.

We note that the centerline temperature 120C lies within the selected time interval. The cursor is moved along the graph until a temperature closest to 120C is determined. Figure below shows that the closest temperature to 120C is 120.014C with the corresponding time t=285.906 sec.

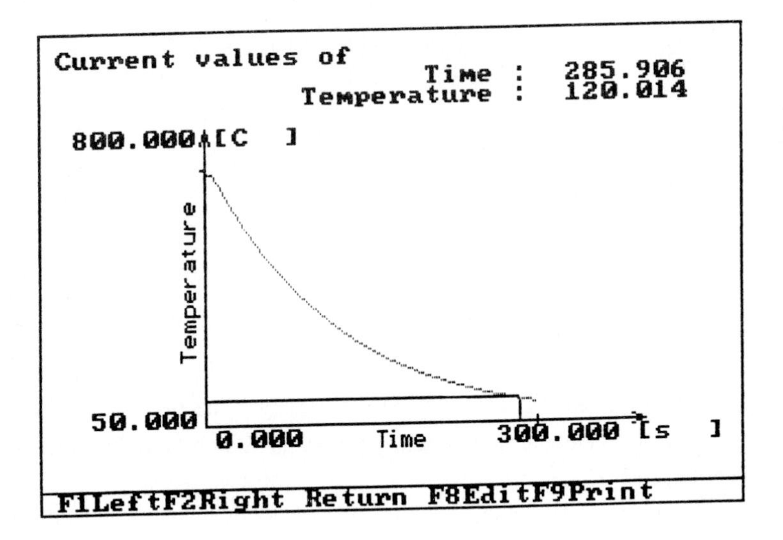

This result can be further refined if a smaller time interval is chosen and the calculations are repeated. This can be done by reducing the time interval, say, from t=275 to t=300 seconds and then repeating the calculations.

Example 4-8. An orange of diameter 10 cm, initially at a uniform temperature T_0=30C, is placed in a refrigerator in which the air temperature is 2C. If the heat transfer coefficient between the air and the surface of the orange is h=50 $W/m^2 \cdot C$, calculate the center temperature of the orange at the time one and a half hours after the start of cooling. Assume thermal properties of orange the same as those for water at the same temperature (i.e., $\alpha=1.4\times10^{-7}$ m^2/s and k=0.59 W/m·K). Figure below shows the geometry.

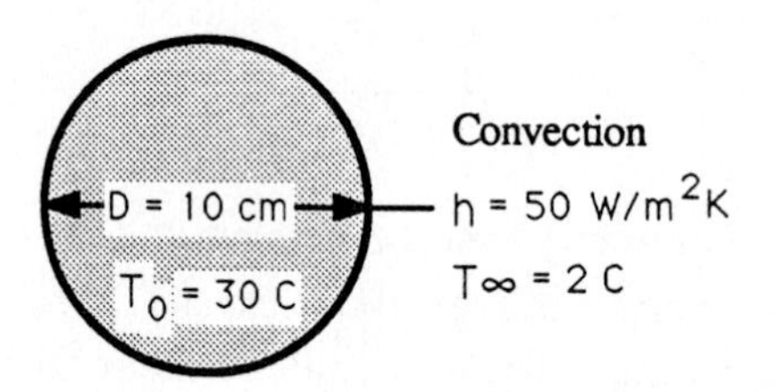

Solution. Table below shows the computer INPUT and OUTPUT data after the program has been run.

TRANSIENT CONDUCTION

Geometry	: Sphere	
Initial temperature	: C	30.000
Surrounding temperature	: C	2.000
Heat transfer coefficient	: $W/m^2 \cdot K$	50.000
Radius	: cm	5.000
Thermal conductivity	: $W/m \cdot K$	0.590
Thermal diffusivity	: m^2/s	0.00000014
Distance from the surface	: cm	5.000
Time	: min	90.000
Temperature	= C	9.495

All the INPUT data are entered as described previously. The last item is the computer OUTPUT which gives the center temperature to be 9.495C at a time 90 minutes after the start of the cooling.

Chapter

FIVE

FINS

Heat transfer by convection between a surface and the surrounding fluid can be increased significantly by attacking the surface thin metal strips of high thermal conductivity, called *fins*. A variety of fin geometries are available for heat transfer applications. For example, in a car radiator the outer surface of the radiator tube is finned because the heat transfer coefficient for air at the outer surface is much smaller than that for water flow at the inner surface. The problem of determination of heat flow through a fin requires a knowledge of temperature distribution in the fin. Heat transfer analysis has been performed for a variety of fin geometries by Harper and Brown (1922), Gardner (1945), Kern and Kraus (1972) and Mikhailov and Ozisik (1984) and analytic expression have been developed for temperature distribution and heat flow rate through the fin. Also, the heat transfer results have been related to a quantity called fin efficiency in order to facilitate the heat transfer calculations in engineering applications. In this Chapter we examine seven different types of fin profiles. For each of these fins, analytic expressions are presented for temperature distribution along the fin and the fin efficiency. The program *FINS* uses these expressions for calculating the fin efficiency and the temperature distribution. Graphical support is provided for determining the temperature distribution along the fin.

5-1 ANALYSIS

In practical applications, heat transfer rate through a fin is related to quantity called fin efficiency, η, defined by

$$\eta = \frac{\text{Actual heat transfer through fin}}{\text{Ideal heat transfer through fin if the entire fin surface were at fin-base temperature } T_b} \qquad (5\text{-}1)$$

Once the fin efficiency is known, heat transfer rate Q_{fin} through a single fin is determined from

$$Q_{fin} = \eta \, Q_{ideal} \qquad (5\text{-}2)$$

where Q_{ideal} is the ideal heat transfer through the fin if the entire fin surface were at fin base temperature T_b.

We present below analytic expressions for the temperature distribution along the fin and the fin efficiency for each of the seven different fin profiles shown in Fig. 5-1, utilizing the results of analysis given by Mikhailov and Ozisik (1984). For all of these fin configurations, the origin of the coordinate axis is taken at the fin base.

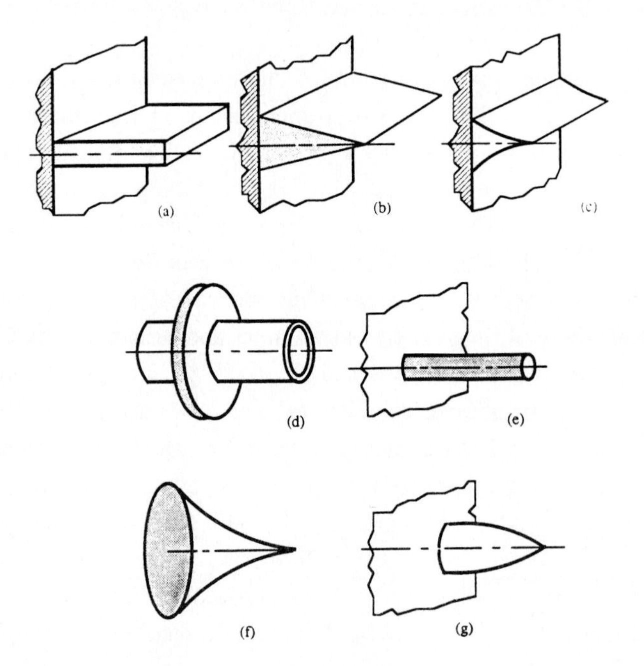

Figure 5-1 Typical fin profiles: (a) longitudinal fin of rectangular profile; (b) longitudinal fin of triangular profile; (c) longitudinal fin of concave parabolic profile; (d) radial fin of rectangular profile; (e) cylindrical spine; (f) spine of concave parabolic profile; (g) spine of convex parabolic profile.

Longitudinal Fin of Rectangular Profile

We consider a longitudinal fin of rectangular geometry having depth L, height b, and base thickness δ_o as illustrated in the following figure. The coordinate axis x is measured from the fin tip. The temperature distribution along the fin is given by

$$T_{fin}(X) = T_\infty + (T_b - T_\infty)\frac{\cosh[M(1-X)]}{\cosh M} \tag{5-3a}$$

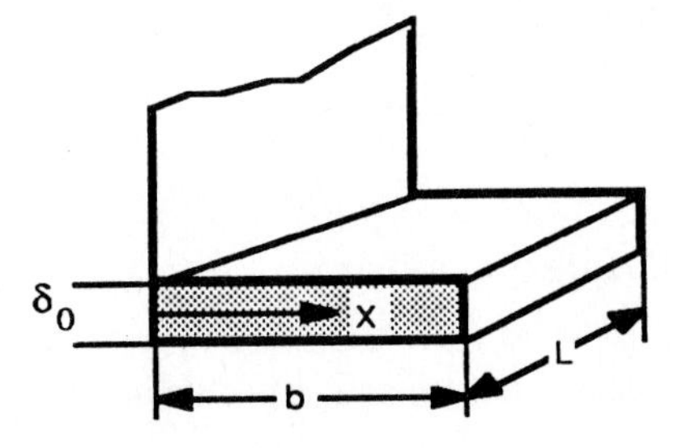

and the fin efficiency η is determined from

$$\eta = \frac{\tanh M}{M} \tag{5-3b}$$

where

T_b = specified fin base temperature
T_∞ = temperature of surrounding fluid

and M and X are defined by

$$M = \left(\frac{2h_{av}}{k\delta_o}\right)^{1/2} \cdot b \quad \text{and} \quad X = \frac{x}{b} \tag{5-4a,b}$$

In addition

b = fin height
h_{av} = average heat transfer coefficient
k = thermal conductivity of fin
δ_o = fin base thickness

and Q_{ideal} is determined as

$$Q_{ideal} = (h_{av})(\text{surface area of fin})(T_b - T_\infty) \tag{5-5a}$$

where

surface area of the fin $\cong$ 2Lb
L = fin depth

Then, Q_{ideal} becomes

$$Q_{ideal} = h_{av}\, 2bL\, (T_b - T_\infty) \tag{5-5b}$$

The efficiency formula given by Eq. (5-3b) has been developed by assuming that the fin tip is insulated, hence does not allow for heat losses from the fin tip. The effect of end losses may be important for very short fins. This may be compensated approximately by using a corrected fin height b_c defined as

$$b_c = b + \frac{1}{2}\delta_o \tag{5-5c}$$

and the resulting error will be less than about 8 percent when

$$\left(\frac{h_{av}\,\delta_o}{2k}\right) \leq \frac{1}{2} \tag{5-5d}$$

Longitudinal Fin of Triangular Profile

Figure below shows a longitudinal fin of triangular profile having depth L, height b and base thickness δ_o.

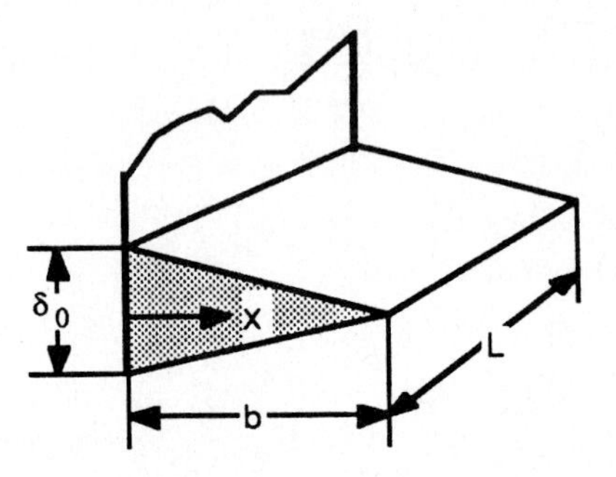

The coordinate axis x is measured from the fin tip. The temperature distribution along the fin is given by

$$T_{fin}(x) = T_\infty + (T_b - T_\infty)\frac{I_o(2M\sqrt{1-X})}{I_o(2M)} \tag{5-6}$$

and the fin efficiency becomes

$$\eta = \frac{I_1(2M)}{M\,I_o(2M)} \tag{5-7}$$

and M and X are defined by Eqs. (5-4a,b). In addition, T_b and T_∞ are, respectively, the fin base and surrounding fluid temperatures, and Q_{ideal} is defined by Eq. (5-5b). I_0 and I_1 are the modified Bessel functions.

Longitudinal Fin of Concave Parabolic Profile

Figure below shows a longitudinal fin of concave parabolic profile having depth L, height b and base thickness δ_o. The origin of the coordinate axis x is taken at the fin tip.

The temperature distribution in the fin is given by

$$T_{fin}(X) = T_\infty + (T_b - T_\infty)(1-X)^{-1/2+\sqrt{(1/4)+M^2}} \tag{5-8}$$

The fin efficiency η is determined according to

$$\eta = \frac{2}{1 + \sqrt{1 + (2M)^2}} \tag{5-9}$$

where M and X are defined by Eqs. (5-4a,b) and Q_{ideal} by Eq. (5-5b).

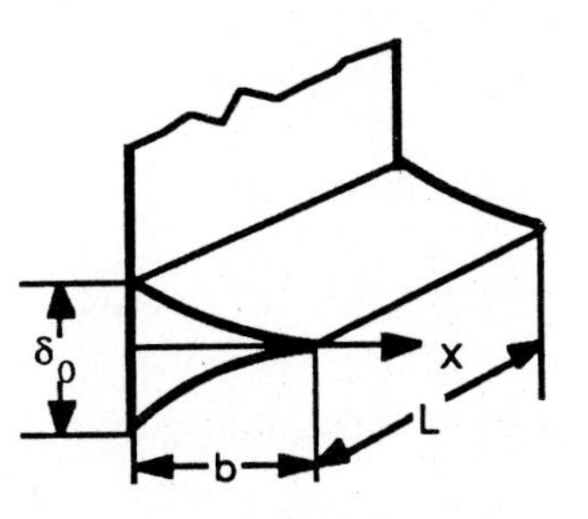

Radial Fin of Rectangular Profile

We now consider radial fins in the form of a circular disc having constant cross-section as illustrated below. The inner and outer radius is, respectively, r_b and r_t, the fin thickness is δ_o.

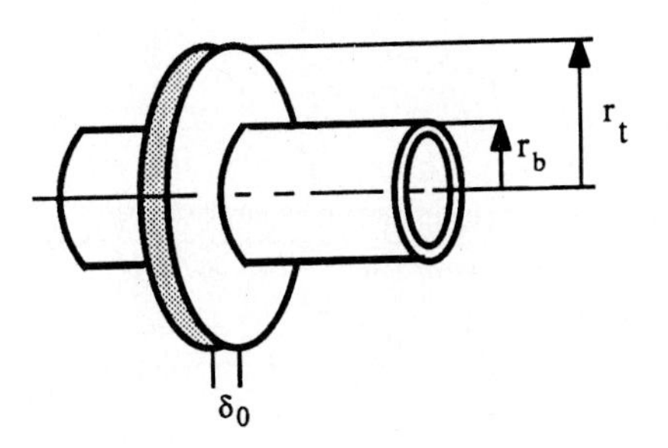

The temperature distribution is determined from

$$T_{fin}(R) = T_\infty + (T_b - T_\infty)\,\frac{K_1(M)\,I_o(MR) + I_1(M)\,K_o(MR)}{K_1(M)\,I_o(MR_b) + I_1(M)\,K_o(MR_b)} \tag{5-10}$$

where

$$R = \frac{r}{r_t}, \qquad R_b = \frac{r_b}{r_t} \tag{5-11a}$$

$$M = (2h_{av}/k\delta_o)^{1/2}\, r_t \tag{5-11b}$$

and I_o, I_1, K_o, and K_1 are the modified Bessel functions. Clearly, $T_{fin}(R_b)=T_b$. The fin efficiency η for radial fins is expressed in the form

$$\eta = \frac{2}{M} \frac{R_b}{1 - R_b^2} \frac{I_1(M)\, K_1(MR_b) - K_1(M) I_1(MR_b)}{I_1(M)\, K_o(MR_b) + K_1(M) I_o(MR_b)} \quad (5\text{-}12a)$$

where R, R_b are defined by

$$R = \frac{r}{r_t}, \qquad R_b = \frac{r_b}{r_t} \quad (5\text{-}12b,c)$$

the ideal heat transfer rate through a radial fin is given by

$$Q_{ideal} = 2\pi(r_t^2 - r_b^2)(T_b - T_\infty)h_{av} \quad (5\text{-}13)$$

and the total heat transfer Q through the fin is determined from

$$Q_{fin} = \eta\, Q_{ideal} \quad (5\text{-}14)$$

Cylindrical Spine

Figure below illustrates a cylindrical spine of height b, radius r_b or the base diameter $\delta_o = 2r_b$. The coordinate axis x is measured from the fin tip.

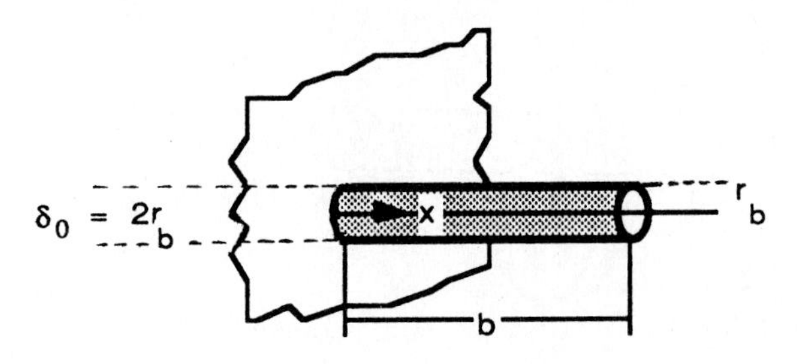

The temperature distribution $T_{fin}(x)$ in the fin is determined from

$$T_{fin}(X) = T_\infty + (T_b - T_\infty) \frac{\cosh[M(1\text{-}X)]}{\cosh M} \quad (5\text{-}15)$$

and the fin efficiency η is given by and

$$\eta = \frac{\tanh M}{M} \quad (5\text{-}16)$$

where M and X are defined by

$$M = \left(\frac{2h_{av}}{kr_b}\right)^{1/2} b, \qquad X = \frac{x}{b} \quad (5\text{-}17a,b)$$

and Q_{ideal} is given by

$$Q_{ideal} = 2\pi\, r_b\, b\, h_{av}(T_b - T_\infty) \quad (5\text{-}17c)$$

Spine of Concave Parabolic Profile

Figure below shows a spine of concave parabolic profile having a height b and base radius δ_0. The origin of the coordinate axis x is taken at the fin tip.

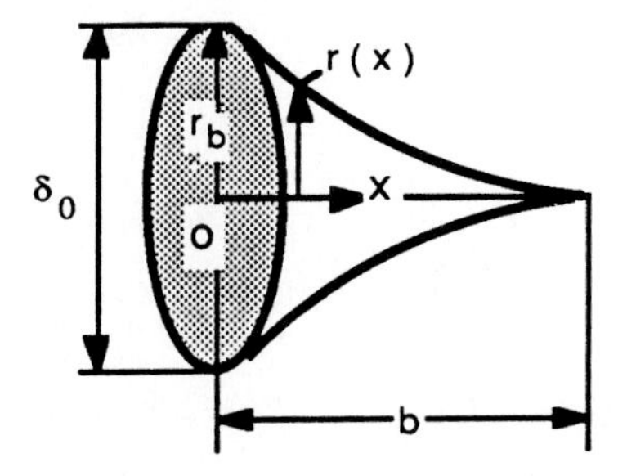

The temperature distribution in the fin, $T_{fin}(x)$, is obtained from

$$T_{fin}(X) = T_\infty + (T_b - T_\infty)(1 - X)^{-3/2+\sqrt{(9/4)+M^2}} \tag{5-18}$$

Clearly, $T_{fin}(0)=T_b$ at X=0. The efficiency of a concave parabolic fin is determined from

$$\eta = \frac{2}{1 + \sqrt{1 + (4/9)M^2}} \tag{5-19}$$

where M and X are defined by

$$M = \left(\frac{2h_{av}}{kr_b}\right)^{1/2} b, \qquad X = \frac{x}{b} \tag{5-20a,b}$$

and Q_{ideal} is given by

$$Q_{ideal} = 2\pi\, r_b\, b\, h_{av}(T_b - T_\infty)\frac{1}{3} \tag{5-20c}$$

Spine of Convex Parabolic Profile

The spine of convex parabolic profile is illustrated in figure below.

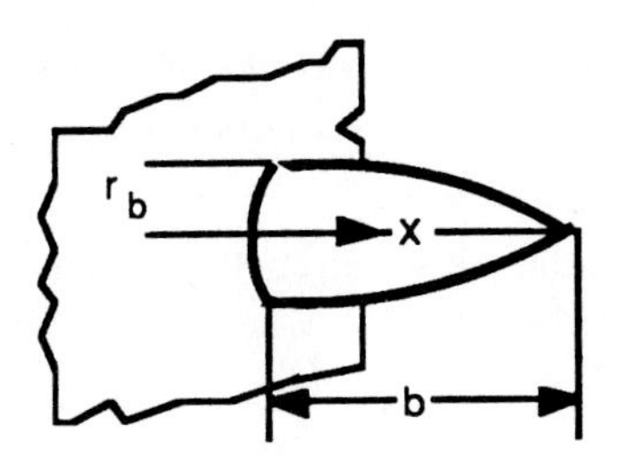

The temperature distribution in the fin is given by

$$T_{fin}(X) = T_\infty + (T_b - T_\infty)\frac{I_0[\frac{4}{3}M(1-X)^{3/4}]}{I_0(\frac{4}{3}M)} \quad (5\text{-}21)$$

Clearly, $T_{fin}(0)=T_b$. The fin efficiency η is determined from

$$\eta = \frac{3}{2}\frac{1}{M}\frac{I_1(\frac{4}{3}M)}{I_0(\frac{4}{3}M)} \quad (5\text{-}22)$$

where I_0 and I_1 are modified Bessel functions, M is defined by

$$M = \left(\frac{2h_{av}}{kr_b}\right)^{1/2} b \quad (5\text{-}23)$$

and Q_{ideal} is given by

$$Q_{ideal} = 2\pi\, r_b\, b\, h_{av}(T_b - T_\infty)\frac{2}{3} \quad (5\text{-}24)$$

Note that Q_{ideal} for fins of concave and convex parabolic profile involve factor $\frac{1}{3}$ and $\frac{2}{3}$, respectively.

5-2 COMPUTER SOLUTIONS

The analytic expressions presented previously are used in the program FINS to calculate the fin efficiency, the heat flow rate through a single fin and the temperature distribution along the fin for each of the seven different fin profiles considered previously. We now illustrate the application with representative examples. In the first example, a detailed description is given of entering the INPUT data into the program, using the graphical support and obtaining tabulated values of temperature distribution along the fin.

Example 5-1. Consider longitudinal copper fin of rectangular profile having a thickness δ_0=1 mm, height b=10 mm and thermal conductivity k=380 W/m·K. The fin base is maintained at a temperature T_b=230C. Fin dissipates heat by convection into the surrounding air at T_∞=30C with a heat transfer coefficient h_{av}=40 W/m^2·C. Figure below shows the geometry. Calculate the fin efficiency η and heat loss per fin for a fin depth L=1 m.

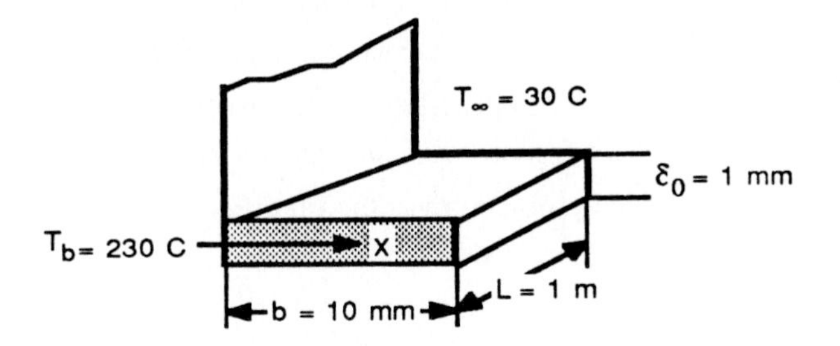

Solution. The following Table shows the computer INPUT and OUTPUT data for the solution of this problem after the program has been run. We describe the main features of entering various INPUT data, the use of graphical support for temperature distribution and obtaining tabulated values of temperature at different locations along the fin.

FINS

Geometry	:	Longitudinal	
Fin profile	:	Rectangular	
Base thickness	:	mm	1.000
Height	:	mm	10.000
Depth	:	m	1.000
Thermal conductivity	:	W/m•K	380.000
Heat transfer coefficient	:	W/m²•K	40.000
Base temperature	:	C	230.000
Surrounding temperature	:	C	30.000
Fin efficiency	=		0.993
Heat loss per fin	=	W	158.887

Entering the INPUT data. The first INPUT data in Fig. 5-9b is the selection of the fin geometry. Move the cursor next to the *Geometry* and press the SPACEBAR. Each time the SPACEBAR is pressed, the geometry changes, successively, to *Longitudinal*, *Radial* and *Spine*. Press the ENTER key when the desired geometry appears on the screen. For the present example it is *Longitudinal* fin.

The next item is the *Fin profile*. To select the fin profile, move the cursor next to the *Fin profile* and press the SPACEBAR. Each time the SPACEBAR is pressed the profile changes, successively, to *Concave parabolic, Triangular*, and *Rectangular*. Press the ENTER key when the desired profile is displayed. In the present example the profile is rectangular.

The next three items are the INPUT data for the fin *Base thickness*, fin *Height* and fin *Depth*. First the UNIT and then the value are entered for each of these three items.

The INPUT data are completed after entering the *Thermal conductivity, Heat Transfer coefficient*, the *Base thickness* and *Surrounding temperatures*.

The solution of the problem gives the fin efficiency $\eta=0.993$ and the heat loss per meter depth of the fin as Q=158.887W.

The last two items are for the computer OUTPUT, which give the *Fin efficiency* and *Heat loss per fin*. The program allows for setting the units of W, kW, cal/s, kcal/h and Btu/hr. In this example it is set to W.

Graphical Support. The graphical support is useful to determine the temperature distribution along the fin. Press the F8 function key to activate the graphical support.

The screen displays the coordinate axis *Temperature* vs. *Distance*. Computer automatically sets the unit and the ranges of temperature and the distance based on the values specified in the example, i.e., $30 \leq T \leq 230$C and $0 \leq x \leq 10$ mm. Press the F8 function key to plot the results. In this particular problem, the curve is very flat since the computer would select the temperature range as $20 \leq T \leq 230$C. For this reason the lower bound for temperature on the graph is changed from 30C to 227C and then the F8 function key is pressed to plot the results over this new range of temperature. The resulting figure is shown below.

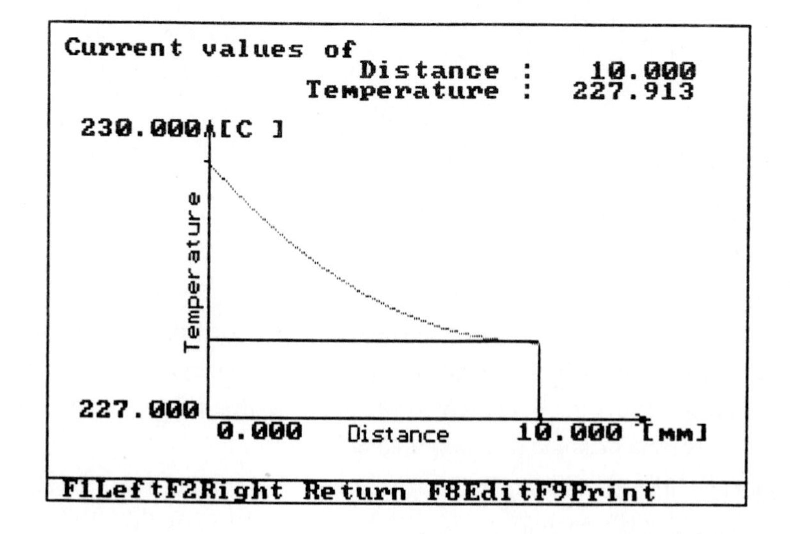

Tabulation of Temperature

If the cursor is moved along the graph, the current values of temperature and location are automatically displayed.

Each time the ENTER key is pressed, the *Distance* and *Temperature* corresponding to the cursor location are stored in the computer memory. Up to 150 such values can be stored, since the computer divides the entire interval 150 equal parts and calculates the temperature at each one of these locations. Therefore, any number of these data can be stored and a hard copy of these results can be obtained by pressing the F9 function key. Table below shows a tabulation of such results for the example considered here.

```
FINS

Geometry    : Longitudinal
Fin profil  : Rectangular
```

Distance [mm]	Temperature [C]
0.000	230.000
1.007	229.600
2.013	229.243
3.020	228.929
4.027	228.657
5.034	228.427
6.040	228.240
7.047	228.095
8.054	227.992

Example 5-2. Repeat Example 5-1 by using the corrected fin height b_c as defined by Eq. (5-5c), i.e.,

$$b_c = b + \frac{1}{2}\delta_o$$

and give only the computer INPUT and OUTPUT data after the program has been run.

Solution. The corrected fin height can be used if the criteria given by Eq. (5-5d) is satisfied, that is

$$\left(\frac{h_{av}\delta_o}{2k}\right) \leq \frac{1}{2}$$

or

$$\frac{40 \times 10^{-3}}{2 \times 380} < \frac{1}{2}$$

Thus the criteria is satisfied and the corrected fin height b_c becomes

$$b_c = b + \frac{1}{2}\delta_o = 10 + \frac{1}{2} = 10.5 \text{ mm}$$

Then, the solution of this problem is exactly the same as that of Example 5-1, except the fin height is taken b_c=10.5 mm instead of b=10 mm. Table below show the resulting computer INPUT and OUTPUT data for Example 5-2. We note that, the fin efficiency changed from 0.993 to 0.992 and the heat loss per fin increased from 158.887W to 166.712W.

FINS

Geometry	: Longitudinal	
Fin profile	: Rectangular	
Base thickness	: mm	1.000
Height	: mm	10.500
Depth	: m	1.000
Thermal conductivity	: W/m•K	380.000
Heat transfer coefficient	: W/m²•K	40.000
Base temperature	: C	230.000
Surrounding temperature	: C	30.000
Fin efficiency	=	0.992
Heat loss per fin	= W	166.712

Example 5-3. Aluminum radial fins of rectangular profile are attached on a circular tube. The inside and outside radius of the circular fin are r_b=1.25 cm and r_t=2.75 cm respectively, the fin thickness is δ_o=1 mm and the thermal conductivity is k=200 W/(m·°C). The tube wall is maintained at T_b=190°C, the surrounding air is at T_∞=40°C and the heat transfer coefficient is h_{av}=80 W/(m^2·°C). Figure below shows the fin geometry. Calculate the fin efficiency and heat loss per fin.

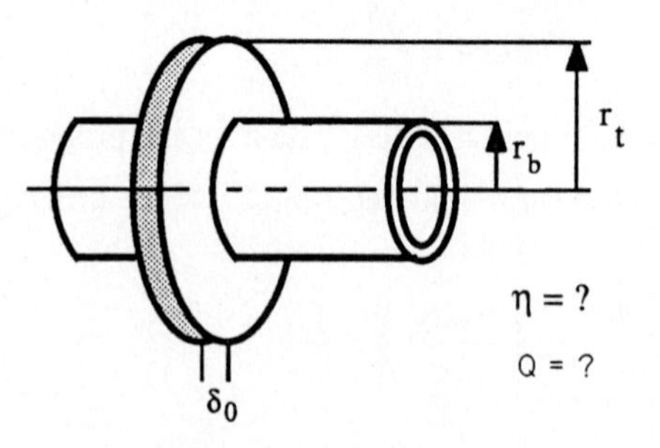

Solution. Table below shows the computer INPUT and OUTPUT data after the program has been run.

```
                         FINS

              Geometry : Radial
           Fin profile : Rectangular

        Base thickness : mm                    1.000
                Height : mm                   15.000
       Fin base radius : cm                    1.250
  Thermal conductivity : W/m•K               200.000
Heat transfer coefficient : W/m²•K            80.000
      Base temperature : C                   190.000
Surrounding temperature : C                   40.000


        Fin efficiency =                       0.919
     Heat loss per fin = W                    41.565
```

For this example, the geometry is set to RADIAL, the fin profile to RECTANGULAR, the fin base thickness is taken as δ_0=1 mm, and the fin height is determined as

$$\text{fin height} = r_t - r_b = 2.75 - 1.25 = 1.50 \text{ cm}$$

and the fin base radius is 1.250 cm. The remaining four INPUT data are straightforward to enter.

The computer OUTPUT gives the fin efficiency as η=0.919 and the heat loss per fin as Q=41.565W.

Example 5-4. Aluminum fins of triangular profile are attached on a plane wall maintained at T_b=240°C. The fin base is δ_0=2 mm, the fin height b=8 mm and the thermal conductivity is k=200 W/(m·°C). The ambient air is at T_∞=40°C and the heat transfer coefficient is h_{av}=50 W/(m^2·°C). Figure below shows the geometry. Calculate the fin efficiency η and the heat loss per fin for a fin depth L=1 m.

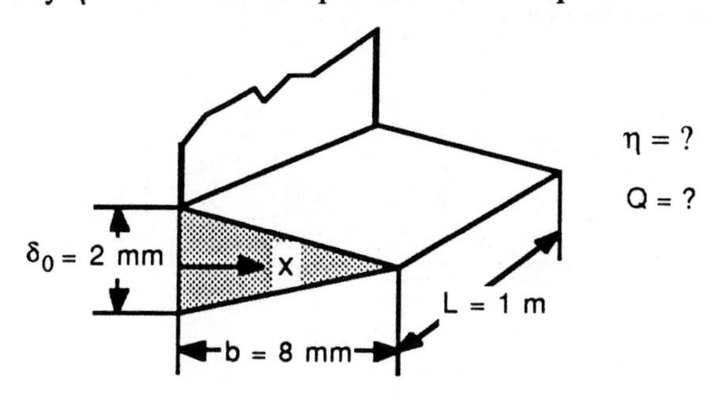

Solution. Table below shows the computer INPUT and OUTPUT data for this example after the program has been run.

```
                         FINS

            Geometry : Longitudinal
         Fin profile : Triangular

      Base thickness : mm                     2.000
              Height : mm                     8.000
               Depth : m                      1.000
Thermal conductivity : W/m•K                200.000
Heat transfer coefficient : W/m²•K           50.000
    Base temperature : C                    240.000
Surrounding temperature : C                  40.000


      Fin efficiency =                        0.992
   Heat loss per fin = W                    158.734
```

The entering the INPUT data is a straightforward matter. The computer output shows that the fin efficiency is η=0.992 and the heat loss per fin is Q=158.734 W.

Example 5-5. This example is intended to show that with increasing fin length, the fin efficiency decreases which in turn results in diminishing returns in heat transfer rate from the fin.

A cylindrical copper spine of radius r_b=0.5 cm, thermal conductivity k=380 W/m·K is attached to a wall maintained at a constant temperature T_b=200C. The surrounding air is at a temperature T_∞=30C and the heat transfer coefficient between the fluid and the spine is h_{av}=15 W/m·K. Figure below shows the geometry. Calculate the fin efficiency and heat loss per fin for the spine length 10 cm, 50 cm, 100 cm, and 1000 cm and compare the heat transfer rates.

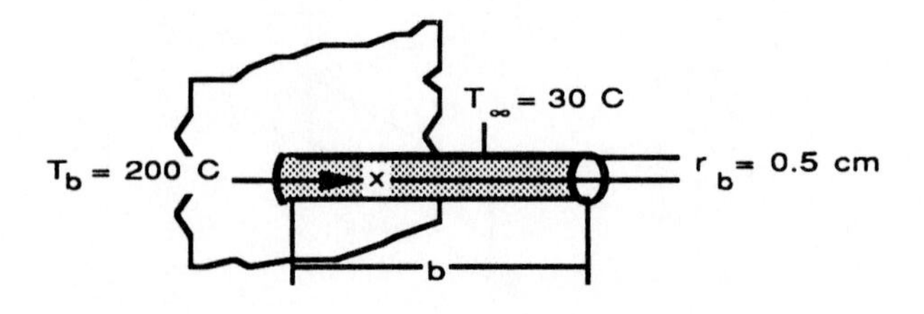

Solution. Table below shows the computer INPUT and OUTPUT data for this example for a spine length b=10 cm. Entering the input data for this example is a straightforward matter.

```
                              FINS

                  Geometry : Spine
               Fin profile : Cylindrical

              Spine radius : cm                         0.500
                    Height : cm                        10.000
      Thermal conductivity : W/m•K                    380.000
 Heat transfer coefficient : W/m²•K                    15.000
          Base temperature : C                        200.000
   Surrounding temperature : C                         30.000


            Fin efficiency =                            0.950
         Heat loss per fin = W                          7.614
```

In Table below we present the fin efficiency and the heat loss per fin for this example for the spine lengths b=10 cm, 50 cm, 100 cm and 1000 cm. Clearly, the fin efficiency reduces drastically which in turn results in diminishing returns in the heat transfer rate from the spine.

Spine height, cm	Spine efficiency, η	Heat loss per spine, W
10	0.950	7.614
50	0.485	19.416
100	0.251	20.146
1000	0.025	20.161

Chapter

SIX

FORCED CONVECTION

So far we considered heat transfer by conduction in solids in which no motion of the medium was involved. In heat transfer by convection, the fluid is in motion and the fluid motion affects the heat transfer. Heat transfer to or from a fluid is said to be by forced convection if the fluid motion is imposed externally by a fan, blower or a pump. Furthermore, the flow may be inside duct or over bodies, or the flow regime may be laminar or turbulent. Therefore, there are so many possibilities of geometry, flow regime, the type of thermal boundary conditions that affect heat transfer in forced convection.

In heat transfer calculations involving forced convection, a quantity of practical interest is the heat transfer coefficient. It is related to the dimensionless quantity, Nusselt number by

$$h = \frac{k \, Nu_m}{D_H}$$

where

k = thermal conductivity of the fluid

Nu_m = Nusselt number

D_H = characteristic length

In this Chapter, recommended analytic expressions are presented for calculating the heat transfer coefficient or the Nusselt number for forced convection inside ducts and flow over bodies for both laminar and turbulent flow. These expressions are used in the program *FORCED CONVECTION* for calculating the heat transfer coefficients.

Once the heat transfer coefficient is known, the heat flow rate between the fluid and wall can be determined from

$$Q = h(T_w - T_f)A$$

where T_w is the wall temperature, T_f is the average fluid temperature and A is the surface area.

6-1 HEAT TRANSFER COEFFICIENTS

We present below various recommended expressions for heat transfer coefficient for forced convection inside ducts and over bodies.

Hydrodynamically Developed Thermally Developing Laminar Flow Inside Ducts

We consider hydrodynamically developed thermally developing laminar flow inside a circular tube of inside diameter "d" or inside a parallel plate duct with a distance "b" between the plates. Hydrodynamically developed flow enters the duct whose walls are maintained at a uniform constant temperature different from the inlet temperature of the fluid. We assume laminar flow, that is the Reynolds number, Re, is less than 2300, i.e.

$$Re = U_m D_H/\nu < 2300 \tag{6-1}$$

where

U_m = mean fluid velocity
ν = kinematic viscosity of fluid

and the hydraulic diameter D_H is defined as

$$D_H = \frac{\begin{pmatrix}4 \times \text{cross section} \\ \text{area for flow}\end{pmatrix}}{\begin{pmatrix}\text{wetted} \\ \text{perimeter}\end{pmatrix}}$$

Then,

D_H = d = tube inside diameter for a circular tube,
D_H = 2b = 2 × spacing between plates for a parallel-plate channel.

The heat transfer between the fluid and the walls of the duct has been originally solved by Graetz analytically for the case of uniform wall temperature and an expression has been developed for the mean heat transfer coefficient h_m or the mean Nusselt number Nu_m over the length from x=0 to x=L of the thermal entry region. The Nu_m is given by

$$Nu_m = \frac{h_m D_H}{k} \tag{6-2}$$

has been related to the dimensionless distance L^* along the duct

$$L^* = \frac{L/D_H}{Re\ Pr} \tag{6-3}$$

where

L = length measured from the beginning of the thermal entry region

D_H = hydraulic diameter

Re = $\frac{U_m D_H}{\nu}$ = Reynolds number

Pr = Prandtl number

k = thermal conductivity of fluid

h_m = mean heat transfer coefficient over the length L.

Table 6-1 gives the values of Nu_m for several different values of the dimensionless duct length L^*. Once L^* is known, the corresponding value of Nu_m is determined by interpolation from the results given in this Table. These results are used to calculate the mean Nusselt number, hence the mean heat transfer coefficient

Table 6-1 Mean Nusselt number Nu_m for a circular tube and a parallel plate channel

L^*	Circular tube	Parallel plate duct
0.000005	93.334	107.833
0.000006	87.769	101.465
0.000007	83.322	96.375
0.000008	79.651	92.173
0.000009	76.545	88.619
0.00001	73.869	85.557
0.000015	64.406	74.728
0.00002	58.429	67.890
0.00003	50.925	59.305
0.00004	46.186	53.885
0.00005	42.813	50.027
0.00006	40.238	47.083
0.00007	38.181	44.731
0.00008	36.483	42.790
0.00009	35.047	41.149
0.0001	33.815	39.736
0.00015	29.442	34.745
0.0002	26.685	31.598
0.0003	23.228	27.657
0.0004	21.049	25.177

Table 6-1 Cont.

L^*	Circular tube	Parallel plate duct
0.0005	19.501	23.416
0.0006	18.321	22.077
0.0007	17.379	21.009
0.0008	16.603	20.130
0.0009	15.948	19.389
0.001	15.384	18.752
0.0015	13.398	16.517
0.002	12.152	15.125
0.003	10.599	13.409
0.004	9.6280	12.354
0.005	8.9432	11.623
0.006	8.4251	11.081
0.007	8.0145	10.662
0.008	7.6783	10.326
0.009	7.3963	10.053
0.01	7.1552	9.8249
0.015	6.3211	9.0972
0.02	5.8146	8.7133
0.03	5.2145	8.3234
0.04	4.8668	8.1277
0.05	4.6406	8.0103
0.06	4.4829	7.9320
0.07	4.3674	7.8761
0.08	4.2796	7.8342
0.09	4.2109	7.8016
0.1	4.1556	7.7766
0.15	3.9895	7.6972
0.2	3.9063	7.6581
1.0	3.657	7.541

Fully Developed Turbulent Flow Inside Ducts

For Reynolds number above about 2300 the flow regime begins to change from laminar to turbulent. Generally, fully developed turbulent flow is established for Reynolds number about above 10^4.

Here we consider hydrodynamically and thermally developed turbulent flow inside ducts. The Nusselt number is given by Notter and Sleicher (1972)

$$Nu = 5 + 0.016\, Re^a\, Pr^b \tag{6-4a}$$

where

$$a = 0.88 - \frac{0.24}{4 + Pr} \quad \text{and} \quad b = 0.33 + 0.5e^{-0.6Pr} \tag{6-4b}$$

which is applicable for

$$0.1 < Pr < 10^4 \tag{6-4c}$$

$$10^4 < Re < 10^6 \tag{6-4d}$$

$$\frac{L}{d} > 25 \tag{6-4e}$$

where d = tube diameter and Re and Nu are defined as

$$Re = \frac{U_m d}{\nu} \tag{6-4f}$$

$$Nu = \frac{hd}{k} \tag{6-4g}$$

The expressions given by Eqs. (6-4) are used in the present software to compute the Nusselt number, hence the heat transfer coefficient h. The program checks the range of correlation for each set of input data and in case the above restrictions on the permissible ranges of Re, Pr and L/D are not satisfied, it prints *Outside correlation range.*

The correlation given by Eqs. (6-4) can be applicable for turbulent flow inside noncircular ducts if the tube diameter d is replaced by the hydraulic diameter D_H. For such cases, case must be exercised in the use of D_H, because with noncircular ducts the heat transfer approaches zero near the sharp corners. Therefore, for certain situations difficulties may arise in applying the circular tube results to a noncircular duct by the hydraulic diameter concept.

Flow Over Bodies

We now consider the mean heat transfer coefficient for flow over bodies such as a flat plate of length L, across a single cylinder or a single sphere of diameter d and a tube bundle.

Flow Over a Flat Plate. Consider flow over a flat plate of length L. The Reynolds number is defined as

$$Re_L = \frac{U_\infty L}{\nu} \tag{6-5}$$

where

L = length of the plate
U_∞ = free-stream velocity
ν = kinematic viscosity of the fluid.

The flow regime begins to change from laminar to turbulent in the range of Reynolds number from 2×10^5 to 5×10^5. Therefore, heat transfer correlations are given separately for the laminar and turbulent flow regimes.

The mean Nusselt number for laminar flow along a flat plate is given by

$$Nu_m = 0.664\, Pr^{1/3}\, Re_L^{1/2} \qquad 0.6 < Pr < 10\,, \quad Re_L \leq 5 \times 10^5 \tag{6-6}$$

$$Nu_m = 0.678\, Pr^{1/3}\, Re_L^{1/2} \qquad Pr \to \infty\,, \quad Re_L \leq 5 \times 10^5 \tag{6-7}$$

where

$$Nu_m = \frac{h_m L}{k} \tag{6-8a}$$

$$Re_L = \frac{U_\infty L}{\nu} \tag{6-8b}$$

and the properties are evaluated at the film temperature. Equation (6-7), for the limiting case of $Pr \to \infty$, is applicable for fluids having large Prandtl number, such as oils.

The mean Nusselt number for turbulent flow along a flat plate of length L is determined from

$$Nu_m = 0.036\, Pr^{0.43}(Re_L^{0.8} - 9200)\left(\frac{\mu_f}{\mu_w}\right)^{0.25} \tag{6-9}$$

for

$$2 \times 10^5 < Re_L < 5.5 \times 10^6 \tag{6-10a}$$

$$0.7 < Pr < 380 \tag{6-10b}$$

$$0.26 < \frac{\mu_f}{\mu_w} < 3.5 \tag{6-10c}$$

where

μ_f = viscosity evaluated at the fluid temperature away from the wall

μ_w = viscosity evaluated at the wall temperature

and all other fluid properties are evaluated at the free stream temperature. For gases, the viscosity correction term $\left(\frac{\mu_f}{\mu_w}\right)^{0.25}$ in Eq. (6-9) is neglected and the physical properties are evaluated at the film temperature $(T_w + T_\infty)/2$. For this particular case, to eliminate the term $(\mu_f/\mu_w)^{0.25}$ from Eq. (6-9) the user can choose $\mu_f = \mu_w$.

Equations (6-6), (6-7) and (6-9) are used in the present software to calculate the mean heat transfer coefficient for laminar and turbulent flow over bodies.

Flow Across a Single Circular Cylinder. We now consider flow across a single cylinder of outside diameter d. The following correlation has been developed by Churchill and Bernstein (1977) for determining the mean heat transfer coefficient h_m for forced convection from gases and liquids to a circular cylinder in cross flow.

$$Nu_m = 0.3 + \frac{0.62\, Re^{1/2}\, Pr^{1/3}}{[1 + (0.4/Pr)^{2/3}]^{1/4}} \left[1 + \left(\frac{Re}{282{,}000}\right)^{5/8}\right]^{4/5} \tag{6-11}$$

for

$$10^2 < Re < 10^7$$

$$Pe = Re \cdot Pr > 0.2$$

Equation (6-11) underpredicts most data by about 20 percent in the range $20{,}000 < Re < 400{,}000$. Therefore, for this particular range of Reynolds number the following modified form of Eq. (6-11) is recommended.

$$Nu_m = 0.3 + \frac{0.62\, Re^{1/2}\, Pr^{1/3}}{[1 + (0.4/Pr)^{2/3}]^{1/4}} \left[1 + \left(\frac{Re}{282{,}000}\right)^{1/2}\right] \tag{6-12}$$

In Eqs. (6-11) and (6-12) all physical properties are evaluated at the film temperature $(T_\infty + T_w)/2$.

The Reynolds and the Nusselt numbers are defined as

$$Re = \frac{U_\infty d}{\nu} \tag{6-13a}$$

$$Nu_m = \frac{h_m d}{k} \tag{6-13b}$$

where

U_∞ = free-stream velocity

d = outside diameter of cylinder

h_m = mean heat transfer coefficient

Equations (6-11) and (6-12) are programmed for determining the mean heat transfer coefficient h_m for flow across a single cylinder. In case the Reynolds and/or Prandtl numbers are out of the intervals specified above, the message *Outside the correlation range* will appear on the screen.

Flow Over a Single Sphere. Consider flow over a single sphere of diameter d. The following correlation has been proposed by Whitaker (1972) for determining the mean heat transfer coefficient h_m for the flow of gases and liquids across a single sphere

$$Nu_m = 2 + (0.4Re^{0.5} + 0.06Re^{2/3})Pr^{0.4}\left(\frac{\mu_\infty}{\mu_w}\right)^{0.25} \tag{6-14}$$

which is valid over the ranges

$$3.5 < Re < 8 \times 10^4$$

$$0.7 < Pr < 380$$

$$1 < \frac{\mu_\infty}{\mu_w} < 3.2$$

and the physical properties are evaluated at the free-stream temperature, except μ_w, which is evaluated at the wall temperature. For gases the viscosity correction is neglected, but the physical properties are evaluated at the film temperature.

The Reynolds and Nusselt numbers are defined as

$$Re = \frac{U_\infty d}{\nu} \tag{6-15a}$$

$$Nu_m = \frac{h_m d}{k} \tag{6-15b}$$

where

U_∞ = the free stream velocity

d = diameter of sphere

h_m = mean heat transfer coefficient.

Equation (6-14) is used in the present software to calculate the mean heat transfer coefficient h_m.

Flow Across Tube Bundles. Heat transfer across tube bundles have numerous applications in the design of heat exchangers and industrial heat transfer equipment. Frequently used tube bundle arrangements include the in-line and the staggered arrangements illustrated in Figs. (6-1a) and (6-1b), respectively. The tube bundle geometry is characterized by the transverse pitch S_T and the longitudinal pitch S_L between the tube centers. To define the Reynolds number for flow through the tube bank, the flow velocity U_{max} is based on the minimum free-flow area available for flow, whether the minimum occurs between the tubes in a transverse row or in a diagonal row. Then the Reynolds number for flow across a tube bank is defined as

$$Re = \frac{D\, G_{max}}{\mu} \tag{6-16a}$$

where

$$G_{max} = \rho U_{max} \tag{6-16b}$$

Here the maximum mass flow velocity G_{max} is the mass flow rate per unit area where the flow velocity is maximum, D is the outside diameter of tube, ρ is the density, and U_{max} is the maximum velocity based on the free-flow area available for fluid flow.

If U_∞ is the flow velocity measured at a point in the heat exchanger before the fluid enters the tube bank (or the flow velocity based on flow inside the heat exchanger shell without the tube bank), the maximum flow velocity U_{max} for in-line arrangement shown in Fig. 5-1a is determined as

$$U_{max} = U_\infty \frac{S_T}{S_T - D} \tag{6-17}$$

where S_T is the transverse pitch and D is the outer diameter of the tube. Clearly, for an in-line arrangement, "S_T–D" is the minimum free-flow area between the adjacent tubes in a transverse row per unit length of the tube.

For the staggered arrangement shown in Fig. 6-1b, the minimum free flow area may occur between adjacent tubes either in a transverse row or in a diagonal row. In the former case, U_{max} is determined from Eq. (6-17); in the latter case it is determined from

$$U_{max} = U_\infty \frac{S_T}{2(S_D - D)} \tag{6-18}$$

Having established the definition of Reynolds number and the maximum flow velocity for flow through a tube bank, we now examine the heat transfer correlations.

Zukauskas (1972) reviewed the works of various investigators and proposed the following correlation for the heat transfer coefficient for flow across tube bundles:

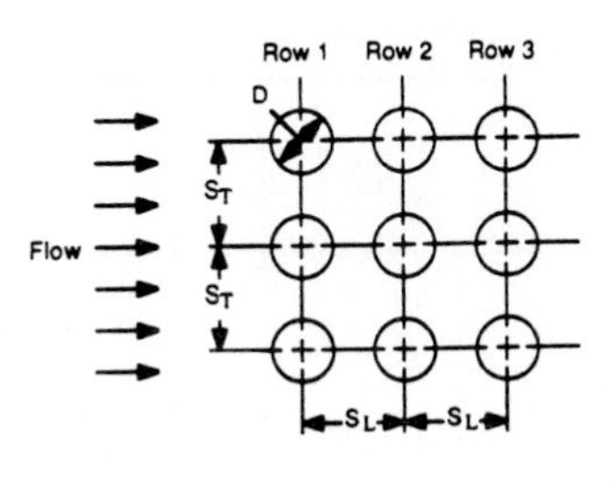

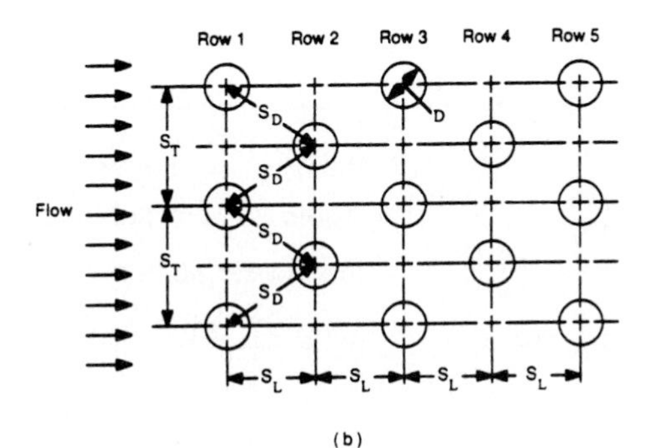

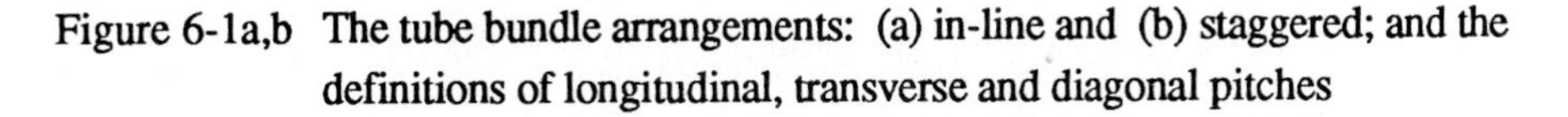
Figure 6-1a,b The tube bundle arrangements: (a) in-line and (b) staggered; and the definitions of longitudinal, transverse and diagonal pitches

$$\frac{h_m D}{k} = c_2 \, Re^m \, Pr^{0.36} \left(\frac{\mu_f}{\mu_w}\right)^n \tag{6-19}$$

where Pr_w is the Prandtl number evaluated at the wall temperature, and

$$n = \begin{cases} 0 & \text{for gases} \\ \frac{1}{4} & \text{for liquids} \end{cases}$$

which is valid for $0.7 < Pr < 500$ and $N \geq 20$. For liquids, the physical properties are evaluated at the bulk mean temperature, since the viscosity correction term is included through the Prandtl number ratio. For gases, the properties are evaluated at the film temperature and the viscosity correction term $(\mu_f/\mu_w)^n$ is taken as unity.

The coefficient c_2 and the exponent m were determined by correlating the experimental data for air, water, and oil reported by numerous investigators. Table 6-2 lists the recommended values of c_2 and m of the correlation given by Eq. (6-19). Equation (6-19) is programmed to compute the heat transfer coefficient h_m; the coefficient c_2 and the exponent m should be obtained from the data in Table 6-2.

Table 6-2 Constant c_2 and exponent m of Eq. (6-19)

Geometry	Re	c_2	m	Remarks
In-line	10 to 10^2	0.8	0.40	
	10^2 to 10^3	Large and moderate longitudinal pitch, can be regarded as a single tube		
	10^3 to 2×10^5	0.27	0.63	
	2×10^5 to 10^6	0.21	0.84	
Staggered	10 to 10^2	0.9	0.40	
	10^2 to 10^3	About 20 percent higher than that for single tube		
	10^3 to 2×10^5	$0.35\left(\frac{S_T}{S_L}\right)^{0.2}$	0.60	$\frac{S_T}{S_L} < 2$
	10^3 to 2×10^5	0.40	0.60	$\frac{S_T}{S_L} > 2$
	2×10^5 to 10^6	0.022	0.84	

Source: Zukauskas (1972).

6-2 COMPUTER SOLUTIONS

The analytic expressions presented previously are used in the program *FORCED CONVECTION* to calculate the average heat transfer coefficient for the following cases:

(1) Hydrodynamically developed, thermally developing laminar flow inside a circular tube and a parallel plate channel

(2) Fully developed turbulent flow inside ducts

(3) Laminar boundary layer flow over a flat plate

(4) Flow over a single cylinder

(5) Flow over a single sphere, and

(6) Flow across tube banks with in-line and staggered tube arrangements.

The following examples illustrate the use of the program *FORCED CONVECTION* to calculate the average heat transfer coefficient for the cases listed above.

The graphic support is also available in order to examine the effects of flow velocity on heat transfer coefficient.

Example 6-1. Engine oil at 40C enters a 2.5 cm-ID, 40 m long tube with a mean flow velocity of 0.7 m/s. The tube wall is maintained at a uniform temperature of 100C. The flow is hydrodynamically developed and thermally developing. Physical properties of engine oil can be taken as k=0.144 W/m·K, ν=0.00024 m^2/s and Pr=2870. Figure below shows the geometry. Calculate the average heat transfer coefficient over the L=40 m length of the tube.

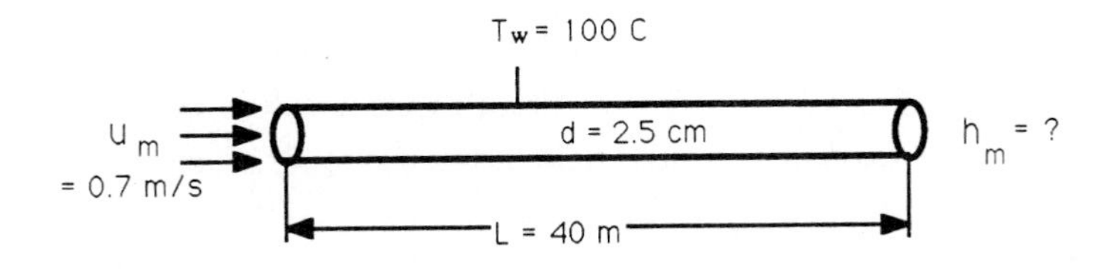

Solution. Table below shows the computer INPUT and OUTPUT data for this problem after the program has been run.

```
                       FORCED CONVECTION

           Type of flow : Inside ducts
               Geometry : Circular tube

               Diameter : cm                         2.500
                 Length : m                         40.000
     Mean flow velocity : m/s                        0.700
Fluid thermal conductivity : W/m•K                   0.14400000
 Fluid kinematic viscosity : m²/s                    0.00024000


        Prandtl number :                          2870.000

       Reynolds number =                              72.917
Heat transfer coefficient = W/m²•K                    44.913
```

The main features of entering the INPUT data in Table above are now described and the use of graphical support is illustrated.

Entering the INPUT data. The first INPUT data is the selection of the type of flow. To set the type of flow, move the cursor next to *Type of flow* press the SPACEBAR. Each time the SPACEBAR is pressed the flow type alternates between flow *Inside ducts* and flow *Over bodies*. Press the ENTER key when the desired flow type is displayed. In this example it is flow *Inside ducts*.

The second input is the *Geometry*. Each time the SPACEBAR is pressed it alternates between *Circular tube* and *Parallel plate channel*. In this example the geometry in circular tube. Therefore, press the ENTER key when *Circular tube* is displayed.

The next 6 items are for the dimensions of the tube, flow velocity and physical properties of the fluid. For each of these items, enter first the unit and then the magnitude.

The last two items are the computer OUTPUT, therefore they are followed by the equal sign "=". The first of these outputs gives the Reynolds number and the second the heat transfer coefficient. For this example, the Reynolds number for the flow is 72.917, thus the flow is laminar, and the heat transfer coefficient is 44.913 $W/m^2 \cdot K$.

Note that, if the input data leads to a result which is outside the correlation range, the computer displays a warning after it has been run. In such a case try another input data.

Graphical Support. The graphical support is available to examine the effects of flow velocity on the heat transfer coefficient. Press the F8 function key to activate the graphical support. The screen displays the coordinate axes *H. T. coefficient* vs. *Flow velocity* computer automatically sets the units and the ranges of the heat transfer coefficient and the flow velocity. The units as well as the ranges of the heat transfer coefficient and the flow velocity can be edited as described previously, thus permitting to zoom to any portion of the curve displayed on the screen. For the heat transfer coefficient any one of the following units can be selected: W/m^2K, kW/m^2K, W/cm^2K, $kcal/m^2hK$ and Btu/ft^2hF. For the velocity built in units are m/s, cm/s, m/h and ft/s.

Figure below show the graphical representation of the solution of Example 6-1.

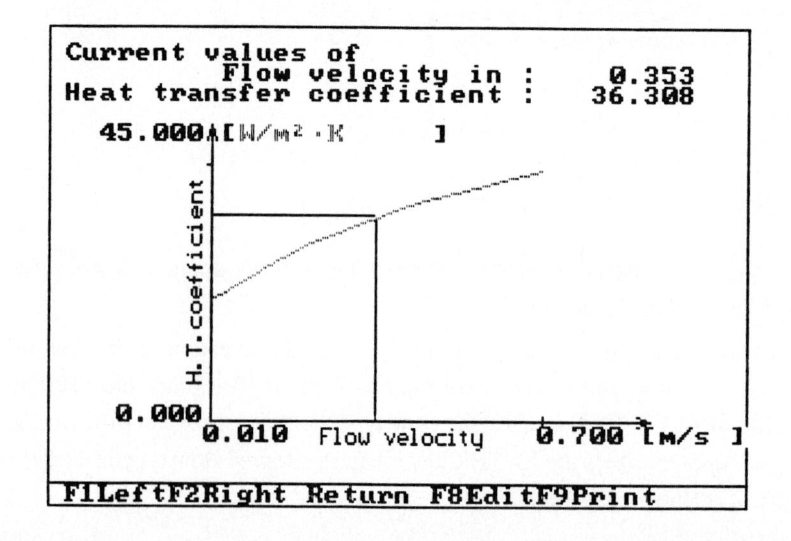

Tabulation of Heat Transfer Coefficients. In the graphical form, each time the ENTER key is pressed, the *Velocity* and *H. T. coefficient* values corresponding to the cursor location are stored in the computer memory and up to 150 such values can be stored. A hard copy of such results stored in the memory can be obtained by pressing the F9 function key. Table below show a tabulation of such results for the Example 6-1.

FORCED CONVECTION

Type of flow : Inside ducts
Geometry : Circular tube

Velocity [m/s]	H.T. coeficient [W/m²•K]
0.010	21.899
0.103	26.534
0.200	31.149
0.302	34.813
0.404	38.082
0.501	40.556
0.603	42.799

Example 6-2. Air at 1 atm and 27C mean temperature flows with a mean velocity of 20 m/s through a 5-cm diameter, 2-m long tube. Flow is hydrodynamically developed and thermally developing. Physical properties are taken as

Thermal conductivity $= k =$ 0.02624 W;/m·K
Kinematic viscosity $= \nu =$ 0.00001846 m^2/s
Prandtl number $=$ 0.708

Figure below shows the geometry. Determine the average heat transfer coefficient over the 2-m length of the tube.

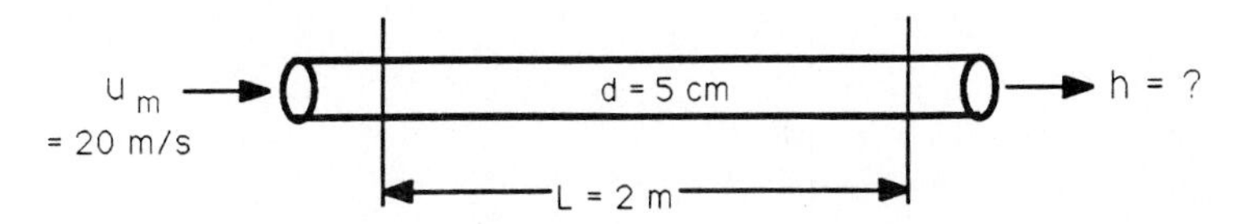

Solution. Table below shows the computer INPUT and OUTPUT data for this example after the program has been run.

```
                    FORCED CONVECTION

              Type of flow : Inside ducts
                  Geometry : Circular tube

                  Diameter : cm                      5.000
                    Length : m                       2.000
       Mean flow velocity : m/s                    20.000
Fluid thermal conductivity : W/m•K                   0.02624000
 Fluid kinematic viscosity : m²/s                    0.00001846


           Prandtl number :                          0.708

          Reynolds number =                      54171.181
  Heat transfer coefficient = W/m²•K                58.858
```

This problem is similar to the one considered in Example 6-1, except the Reynolds number is 54177.181, that is the flow is turbulent, and the resulting heat transfer coefficient is h=58.858 $W/m^2 \cdot K$.

Example 6-3. The thermal insulation is removed from 1-m length section of a steam pipe of outside diameter d=25 cm, carrying high-pressure steam at 180C. Air at -5C is flowing across the exposed section of the pipe with a velocity of U_m = 6 m/s. Physical properties of air are taken as

$k = 0.031$ W/m·K

$\nu = 0.00002184$ m^2/s

$Pr = 0.695$

Figure below shows the geometry. Calculate the average heat transfer coefficient between air and the steam pipe.

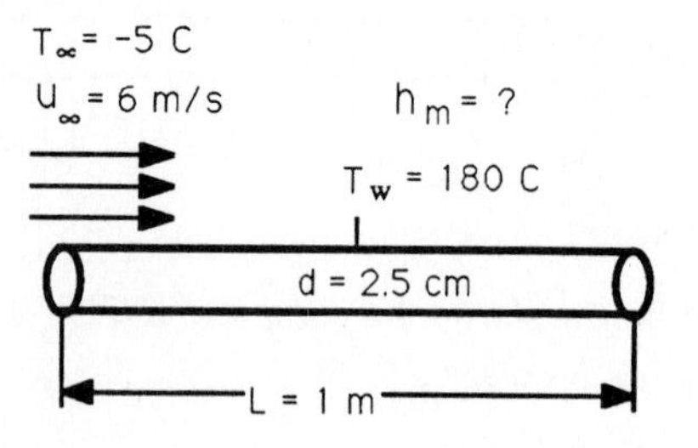

Solution. Table below shows the computer INPUT and OUTPUT data for this example after the program has been run.

```
                    FORCED CONVECTION

               Type of flow : Over bodies
                   Geometry : Single circular cylinder

                   Diameter : cm                   25.000

         Mean flow velocity : m/s                   6.000
 Fluid thermal conductivity : W/m•K                 0.03100000
  Fluid kinematic viscosity : m²/s                  0.00002184


            Prandtl number  :                       0.695

           Reynolds number  =                   68681.319
  Heat transfer coefficient = W/m²•K               23.408
```

In this example the type of flow is *Over bodies*, and the geometry is *Single circular cylinder*. Entering the input data is a straightforward matter. The solution shows that the Reynolds number is 68681.319 and the resulting *Heat transfer coefficient* is 23.408 $W/m^2 \cdot K$.

If the input data violates the range of correlation, the following warning message appears on the screen:

Outside the correlation range

For example, if the cylinder diameter is changed from 25 cm, say, to 40 m, such a warning message will appear.

Example 6-4. Water at 40C flows with a velocity of 2 m/s across a 2.5 cm diameter sphere. The surface of the sphere is maintained at a uniform temperature 100C. Physical properties of water are taken as

Thermal conductivity = k = 0.629 W/m·K

Kinematic viscosity = ν = 0.00000066

Viscosity at 40C = 0.000654 kg/m·s

Viscosity at 100C = 0.000282 kg/m·s

Prandtl number = Pr = 4.340

Figure below shows the geometry. Calculate the average heat transfer coefficient between the water and the sphere.

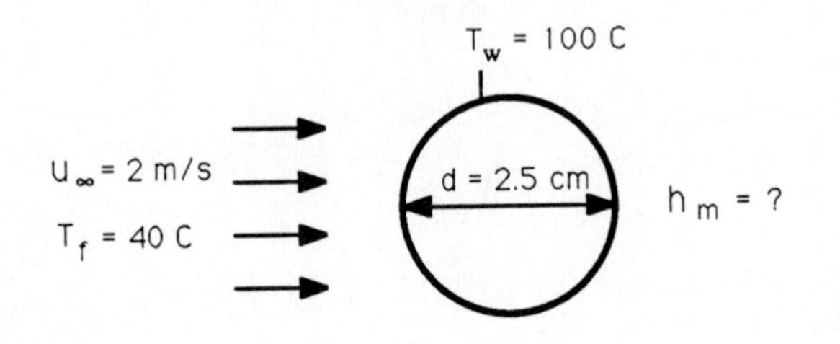

Solution. Table below shows the computer INPUT and OUTPUT data for this problem after the program has been run.

```
FORCED CONVECTION

            Type of flow : Over bodies
                Geometry : Single sphere

                Diameter : cm                  2.500

      Mean flow velocity : m/s                 2.000
Fluid thermal conductivity : W/m•K             0.62900000
 Fluid kinematic viscosity : m²/s              0.00000066
    Fluid viscosity at Tf : kg/m•s             0.00065400
    Fluid viscosity at Tw : kg/m•s             0.00028200

         Prandtl number :                      4.340

        Reynolds number =                    75757.576
 Heat transfer coefficient = W/m²•K          12199.276
```

In this example the geometry is *Single sphere*. The correlation requires a viscosity correction; therefore viscosity of the fluid at both the sphere temperature and the bulk fluid temperatures are given. Entering the INPUT data is a straightforward matter. The OUTPUT data shows that the Reynolds number is Re=75757.576 and the heat transfer coefficient is h_m=12199.276 W/m^2·K.

Example 6-5. Air at atmospheric pressure and 27C flows over a tube bank consisting of D=1 cm diameter tubes 10 rows deep. The flow velocity before the air enters the tube bundle is U_m=1 m/s. Tubes are in equilateral-triangular arrangement with $S_T/D=S_D/D=1.25$ as illustrated in Fig. Ex6-5. Physical properties of air are taken as

Thermal conductivity = k = 0.026 W/m·K

Kinematic viscosity = ν = 0.00001568 m^{2}s

Viscosity at $T_f=\mu_f$ = 0.00001840 kg/m·s

Calculate the average heat transfer coefficient between air and the tube bank.

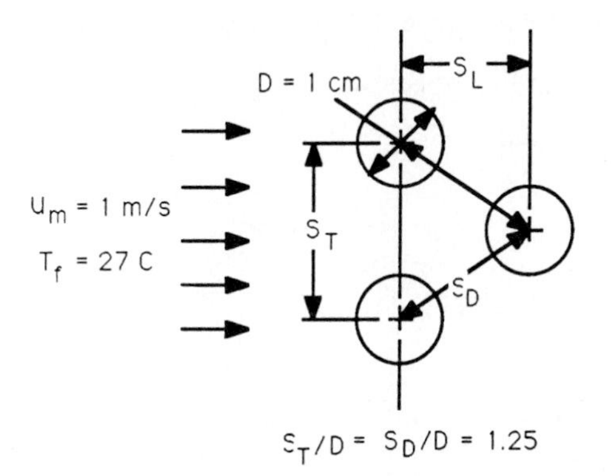

Solution. Table below shows the computer INPUT and OUTPUT data for this example after the program has been run.

```
                    FORCED CONVECTION

           Type of flow : Over bodies
               Geometry : Tube bundles

       Outside diameter : cm                      1.000

     Mean flow velocity : m/s                     1.000
Fluid thermal conductivity : W/m•K                0.02600000
 Fluid kinematic viscosity : m²/s                 0.00001568
     Fluid viscosity at Tf : kg/m•s               0.00001840
     Fluid viscosity at Tw : kg/m•s               0.00001840
   Constants  C2 :  0.360   m :  0.60           n :  1.00
          Prandtl number :                        0.708

         Reynolds number =                        637.755
 Heat transfer coefficient = W/m²•K                39.818
```

For this problem the type of flow is flow *Over bodies* and the geometry is *Tube bundles*. Most of the INPUT data are readily entered. However, the heat transfer correlation for tube banks, as given by Eq. (6-19) involves viscosity correction. Therefore fluid viscosities at bulk temperature and wall temperature are generally needed. For air such correction is not needed; therefore we set $\mu_f = \mu_w = 0.00001840$. The constant c_2 and the exponent m are obtained from Table 6-2 as

$$c_2 = 0.360 \quad \text{and} \quad m = 0.60$$

and the exponent for viscosity correction is set to n=1.

The computer output data gives the Reynolds number as Re=637.755 and the heat transfer coefficient as $h_m = 39.818\ W/m^2{\cdot}K$

Example 6-6. Repeat Example 6-5 for the case of tubes in square arrangement with $\frac{S_T}{D} = \frac{S_L}{D} = 1.25$ as shown in Figure below.

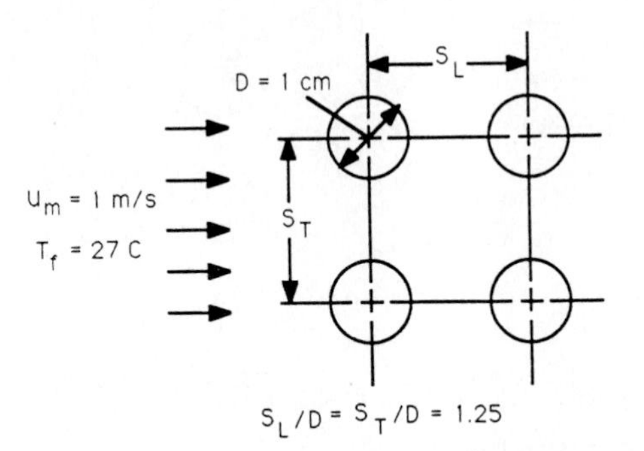

Solution. The problem is similar to that considered in Example 6-5, except for this case the constants c_2 and m are obtained from Table 6-2 for the case of in-line arrangement, to give

$$c_2 = 0.270 \qquad \text{and} \qquad m = 0.63$$

Table below shows that the heat transfer coefficient is h_m=36.248 $W/m^2{\cdot}K$, which is less than that in Example 6-5.

```
                    FORCED CONVECTION

          Type of flow : Over bodies
              Geometry : Tube bundles

      Outside diameter : cm                          1.000

   Mean flow velocity : m/s                          1.000
Fluid thermal conductivity : W/m•K                   0.02600000
 Fluid kinematic viscosity : m²/s                    0.00001568
     Fluid viscosity at Tf : kg/m•s                  0.00001840
     Fluid viscosity at Tw : kg/m•s                  0.00001840
   Constants  C2 :  0.270    m :  0.63             n :  1.00
           Prandtl number :                          0.708

          Reynolds number =                          637.755
 Heat transfer coefficient = W/m²•K                   36.248
```

Chapter

SEVEN

FREE CONVECTION

There are situations in which convective motion sets up in the fluid without a forced velocity generated by a fan, a blower or a pump. Consider for example, a hot plate placed vertically in a body of fluid at rest which is at a uniform temperature lower than the temperature of the plate. The temperature variation within the fluid will generate a density gradient which, in a gravitational field, will give rise, in turn, to a convective motion as a result of buoyancy forces. The fluid motion set up as a result of the buoyancy force is called free convection, or natural convection. In this chapter, we consider the determination of heat transfer coefficient for free convection for the following geometries:

Vertical plate
Horizontal plate - hot surface faces up
Horizontal plate - hot surface faces down
Vertical cylinder
Horizontal single cylinder
Single sphere

First, expressions are presented for the computation of Nusselt number, Nu. Once this quantity is known, then the heat transfer coefficient is determined from

$$h = \frac{k\,Nu}{L}$$

where k is the fluid thermal conductivity and L is the characteristic length. In this Chapter, h and Nu refer to the mean values, i.e., h_m and Nu_m.

7-1 HEAT TRANSFER COEFFICIENTS FOR FREE CONVECTION

We present below various recommended expressions for free convection heat transfer coefficients for the geometric arrangements listed above. In all the expressions given below, the Nusselt number for free convection is correlated as a function of the Prandtl number and the Rayleigh number defined as

$$Ra = \frac{gL^3 \beta |T_w - T_f| Pr}{\nu^2} \tag{7-1}$$

where

β = volumetric thermal expansion coefficient
L = characteristic length
Pr = Prandtl number
T_w = wall temperature
T_f = Fluid temperature
g = gravitational acceleration,
υ = kinematic viscosity

and the Nusselt number is defined as

$$Nu = \frac{hL}{k} \tag{7-2}$$

where

h = heat transfer coefficient
k = thermal conductivity
L = characteristic length

For the present analysis, the characteristic length L for various geometries is defined as

Geometry	Characteristic Length, L
Vertical plate	Plate height
Vertical cylinder	Cylinder height
Horizontal cylinder	Cylinder diameter
Sphere	Sphere diameter
Horizontal plate heated surface up or down	$\frac{\text{Surface area}}{\text{Perimeter}} \equiv \frac{A}{p}$ †

and the physical properties are evaluated at the mean film temperature.

Vertical Plate. Churchill and Chu (1975) proposed the following two equations for correlating free convection on a vertical plate under isothermal surface conditions,

$$Nu = 0.68 + \frac{0.67\, Ra^{1/4}}{[1 + (0.492/Pr)^{9/16}]^{4/9}} \tag{7-3a}$$

which applies only to the laminar flow regime, $10^{-1} < Ra < 10^9$ and holds for all values of the Prandtl number.

† In the computer program the characteristic length is taken as A/P. If any other value is to be used for the characteristic length, say L^*, one need to choose the values of A and P such that the ratio "A/P" equals to L^*.

The other relation given by

$$Nu = \left\{0.825 + \frac{0.837\ Ra^{1/6}}{[1 + (0.492/Pr)^{9/16}]^{8/27}}\right\}^2 \qquad (7\text{-}3b)$$

applies to both laminar and turbulent regions. Here we prefer to use Eq. (7-3a) for the laminar flow regime, $10^{-1} < Ra < 10^9$ and Eq. (7-3b) for the range $10^9 < Ra < 10^{12}$.

Horizontal Plate Hot Surface Facing Up or Cold Surface Facing Down

$$Nu = 0.54\ Ra^{1/4} \quad \text{for} \quad 10^5 < Ra < 2 \times 10^7 \qquad (7\text{-}4a)$$

$$Nu = 0.14\ Ra^{1/3} \quad \text{for} \quad 2 \times 10^7 < Ra < 3 \times 10^{10} \qquad (7\text{-}4b)$$

Horizontal Plate Hot Surface Facing Down or Cold Surface Facing Up

$$Nu = 0.27\ Ra^{1/4} \quad \text{for} \quad 3 \times 10^5 < Ra < 3 \times 10^{10} \qquad (7\text{-}5)$$

Vertical cylinder

For fluids having a Prandtl number 0.7 and higher, a vertical cylinder may be treated as a vertical plate when

$$\frac{L/D}{(Gr)^{1/4}} < 0.025 \qquad (7\text{-}6)$$

where Gr is the Grashof number, (i.e., $Ra = Gr \cdot Pr$), L is the cylinder height and D is the cylinder diameter. If the condition given by Eq. (7-6) is not satisfied, the computer will print on the screen *Outside the correlation range.*

Horizontal Single Cylinder

The following correlation is proposed by Churchill and Chu [1975]

$$Nu = 0.6 + \frac{0.387\ Ra^{1/6}}{[1 + (0.559/Pr)^{9/16}]^{8/27}} \qquad (7\text{-}7)$$

for $10^{-4} < Ra < 10^{12}$

where

$$Nu = \frac{hD}{k}, \qquad Ra = \frac{g\ \beta |T_w - T_f| D^3}{\nu^2} Pr$$

Single Sphere

Amato and Tien [1972] proposed the following correlation

$$Nu = 2 + 0.5\ Ra^{1/4} \tag{7-8}$$

for $3 \times 10^5 < Ra < 8 \times 10^8$

and Nu and Ra numbers are based on the sphere diameter.

7-2 COMPUTER SOLUTIONS

Analytic expressions presented previously are used in the program *FREE CONVECTION* to calculate the average free convection heat transfer coefficient for geometries including:

Vertical plate
Horizontal plate - hot surface facing up
Horizontal plate - hot surface facing down
Vertical cylinder
Horizontal single cylinder, and
Single sphere

For all the problems considered in this chapter, the followings are the common input data:

T_w = Wall temperature
T_f = Fluid temperature
k = Fluid thermal conductivity
ν = Fluid kinematic viscosity
Pr = Prandtl number
β = Volumetric expansion coefficient

The graphic support is also available in order to examine the effects of temperature difference on free convection heat transfer coefficient. The following examples illustrate the use of the program *FREE CONVECTION.*

Example 7-1. A vertical plate 0.3 m high, 1 m wide, maintained at a uniform temperature 124C is exposed to quiescent atmospheric air at 30C. The physical properties of air can be taken as:

$$K = 0.03\ W/m{\cdot}K, \quad \nu = 0.00002076\ m^2/s, \quad Pr = 0.697$$

and the volumetric thermal expansion coefficient, $\beta = 0.00286$ 1/K. Figure below shows the geometry. Calculate the average free convection heat transfer coefficient between the plate and air.

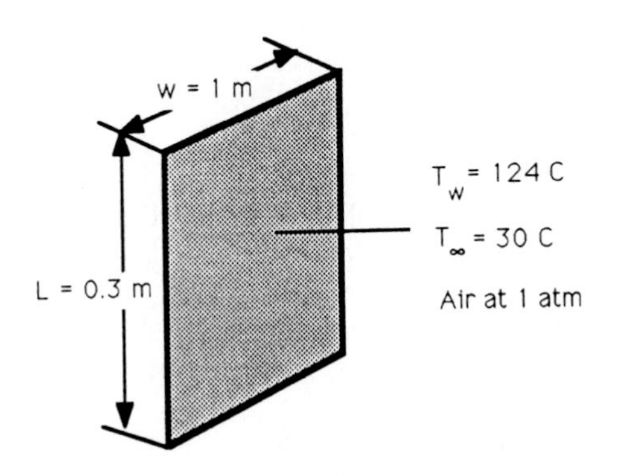

Solution. Table below shows the computer INPUT and OUTPUT data for this example after the program has been run.

```
                         FREE  CONVECTION
              Wall temperature : C                        124.000
             Fluid temperature : C                         30.000
                      Geometry : Vertical Plate
                        Height : m                          0.300

    Fluid thermal conductivity : W/m•K                      0.030
    Fluid  kinematic viscosity : m²/s                       0.000020760
               Prandtl number :                             0.697
Volumetric expansion coefficient: 1/°K                      0.002860000


               Rayleigh number =                       1.1516E+08
     Heat Transfer Coefficient = W/m²•K                     5.384
```

In this program, the Rayleigh number is also an output in addition to the heat transfer coefficient. The graphical representation is given below.

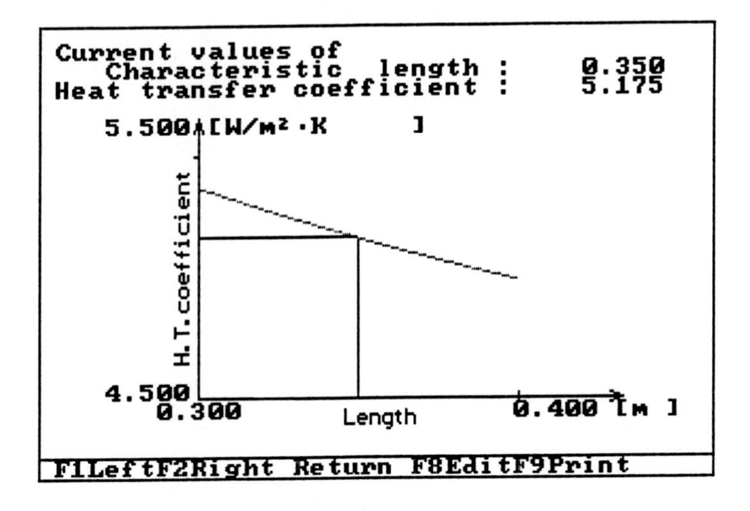

Note that the plate width w is not included in the computer INPUT data, because it cancels out.

Example 7-2. A cube of sides 5 cm is suspended with one of its surfaces in horizontal position in quiescent atmospheric air at 20C. All surfaces of the cube are maintained at a uniform temperature of 100C. Figure below shows the geometry.

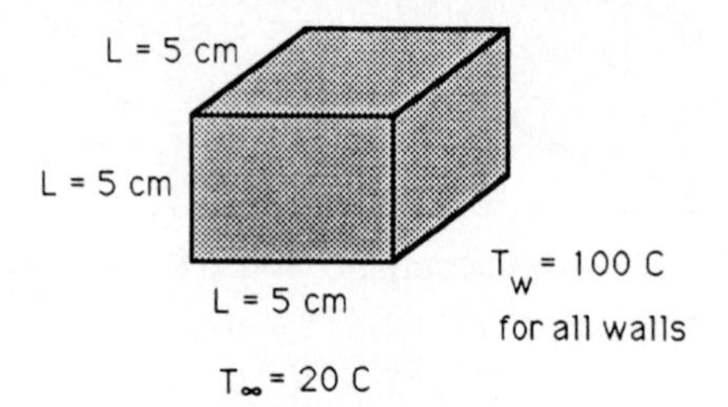

The physical properties of air can be taken as

Thermal conductivity = 0.029 W/m·K
Kinematic viscosity = 0.0000189 m^2/s
Prandtl number = 0.701
Volumetric thermal expansion coef. = 0.003 K^{-1} ,

Calculate the free convection heat transfer coefficient for the (i) Vertical, (ii) top surface (i.e., hot surface facing up) and (iii) Bottom surface (i.e., hot surface facing down). The following three Tables show the input and the output data for the cases of vertical, top and bottom surfaces of the cube, respectively.

```
                    FREE   CONVECTION
          Wall temperature : C                          100.000
         Fluid temperature : C                           20.000
                  Geometry : Vertical Plate
                    Height : cm                           5.000

Fluid thermal conductivity : W/m•K                        0.029
Fluid  kinematic viscosity : m²/s                         0.000018900
            Prandtl number :                              0.701
Volumetric expansion coefficient: 1/°K                    0.003000000

           Rayleigh number =                         577543.462
 Heat Transfer Coefficient = W/m²•K                       8.605
```

```
                    FREE   CONVECTION
           Wall temperature : C                           100.000
          Fluid temperature : C                            20.000
                   Geometry : Horizontal Plate - Surface Up
               Surface area : cm²                          25.000
                  Perimeter : cm                            5.000
 Fluid thermal conductivity : W/m•K                         0.029
 Fluid  kinematic viscosity : m²/s                          0.000018900
             Prandtl number :                               0.701
Volumetric expansion coefficient: 1/°K                      0.003000000

            Rayleigh number =                          577543.462
   Heat Transfer Coefficient = W/m²•K                       8.634
```

```
                    FREE   CONVECTION
           Wall temperature : C                           100.000
          Fluid temperature : C                            20.000
                   Geometry : Horizontal Plate - Surface Down
               Surface area : cm²                          25.000
                  Perimeter : cm                            5.000
 Fluid thermal conductivity : W/m•K                         0.029
 Fluid  kinematic viscosity : m²/s                          0.000018900
             Prandtl number :                               0.701
Volumetric expansion coefficient: 1/°K                      0.003000000

            Rayleigh number =                          577543.462
   Heat Transfer Coefficient = W/m²•K                       4.317
```

For the horizontal surfaces we have chosen the characteristic length L=5 ≡ side of the square by setting A=25 cm^2, P=5 cm so that L ≡ (A/p) = (25/5) = 5 cm.

The entering the input data is straightforward. The solutions for the problem are

h = 8.605 $W/m^2{\cdot}K$ for the vertical surface

h = 8.634 $W/m^2{\cdot}K$ for the top surface

h = 4.317 $W/m^2{\cdot}K$ for the bottom surface

Example 7-3. An electric heater of outside diameter D=2.5 cm and length L=1 m is immersed horizontally inside a large tank containing engine oil at 140C. The fluid properties can be taken as:

Thermal conductivity = 0.138 W/m·K

Kinematic viscosity = 0.0000375 m^2/s

Prandtl number = 300

Volumetric thermal expansion coef. = 0.0007 K^{-1}.

The geometry is shown in Figure below. Calculate the free convection heat transfer coefficient.

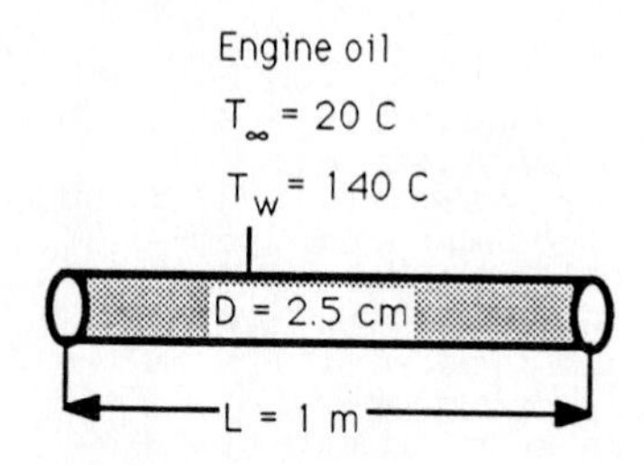

Solution. Table below shows the computer input and output data for this example after the program has been run.

```
                    FREE   CONVECTION
           Wall temperature : C                        140.000
          Fluid temperature : C                         20.000
                   Geometry : Horizontal Cylinder
                   Diameter : cm                         2.500

 Fluid thermal conductivity : W/m•K                      0.138
 Fluid  kinematic viscosity : m²/s                       0.000037500
            Prandtl number :                           300.000
Volumetric expansion coefficient: 1/°K                   0.000700000


            Rayleigh number =                        2.7468E+06
   Heat Transfer Coefficient = W/m²•K                  145.893
```

Example 7-4. A cylindrical electric heater of outside diameter D=2.5 cm and length L=2 m is immersed horizontally into a pool of mercury at 100C. The surface of the heater is kept at an average temperature of 300C. The physical properties of mercury can be taken as

Thermal conductivity = 12.340 W/m·K
Kinematic viscosity = 0.00000008 m^2/s
Prandtl number = 0.012
Volumetric thermal expansion coef. = 0.000182 K^{-1}.

Figure below shows the geometry. Calculate the free convection heat transfer coefficient.

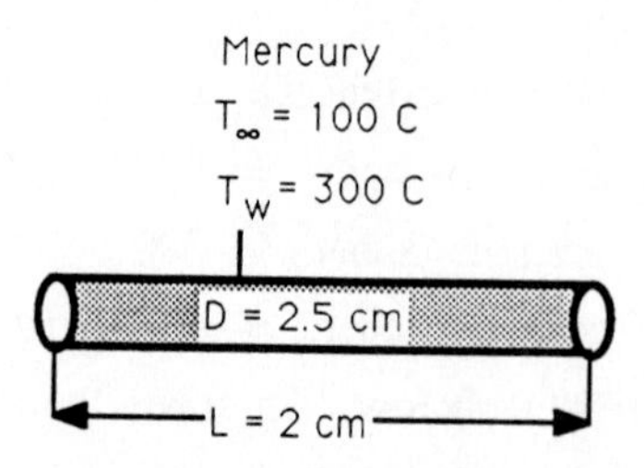

Solution. Table below shows the computer input and output data for this example.

```
                  FREE   CONVECTION
         Wall temperature : C                        300.000
        Fluid temperature : C                        100.000
                 Geometry : Horizontal Cylinder
                 Diameter : cm                         2.500

 Fluid thermal conductivity : W/m•K                   12.340
 Fluid  kinematic viscosity : m²/s                     0.000000080
            Prandtl  number :                          0.012
Volumetric expansion coefficient: 1/°K                 0.000182000


            Rayleigh number =                     1.0461E+07
  Heat Transfer Coefficient = W/m²•K               6120.250
```

Note that the heat transfer coefficient for mercury is about 35 times higher than that for oil considered in Example 7-3 under the same conditions.

Example 7-5. A sphere of diameter D=5 cm is immersed into quiescent water at 30C. The surface of the sphere is kept at 120C. The physical properties of water can be taken as:

Thermal conductivity	=	0.664 W/m·K
Kinematic viscosity	=	0.000000393 m^2/s
Prandtl number	=	2.420
Volumetric thermal expansion coef.	=	0.00018 K^{-1}

Figure below shows the geometry. Calculate the free convection heat transfer coefficient.

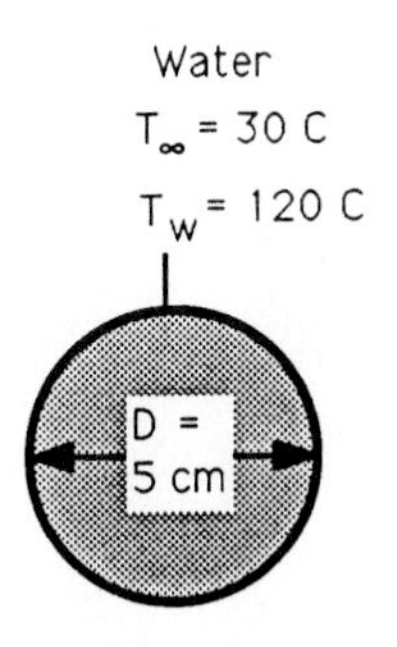

Solution. Table below shows the computer input and output after the program has been run.

```
                      FREE  CONVECTION
             Wall temperature : C                       120.000
            Fluid temperature : C                        30.000
                     Geometry : Sphere
                     Diameter : cm                        5.000

   Fluid thermal conductivity : W/m•K                     0.664
   Fluid  kinematic viscosity : m²/s                      0.000000393
              Prandtl number :                            2.420
Volumetric expansion coefficient: 1/°K                    0.000180000

              Rayleigh number =                       3.1126E+08
    Heat Transfer Coefficient = W/m²•K                  908.521
```

Chapter

EIGHT

FILMWISE CONDENSATION

Condensers constitute an important and widely used type of heat exchanger with unique characteristics of heat transfer mechanism on the condensing and boiling side. If a vapor strikes a surface that is at a temperature below the corresponding saturation temperature, the vapor will immediately condense into the liquid phase. If the condensation takes place continuously over the surface which is kept cooled by some cooling process and the condensed liquid is removed from the surface by the motion resulting from gravity, then the condensing surface is usually covered with a thin layer of liquid, and the situation is known as *FILMWISE CONDENSATION*. The presence of a liquid film over the surface constitutes a thermal resistance to heat flow. Therefore, numerous experimental and theoretical investigations have been conducted to determine the heat transfer coefficient for film condensation over surfaces. The first fundamental analysis leading to the determination of the heat transfer coefficient during filmwise condensation of pure vapors (i.e., without the presence of non-condensable gas) on a flat plate and a circular tube was given by Nusselt in 1916. Over the years, improvements have been made on Nusselt's theory of film condensation. But with the exception of the condensation of liquid metals, Nusselt's original theory has been successful and still is widely used.

In this Chapter we present some of the correlations for the calculation of the average heat transfer coefficients for the filmwise condensation of steam. The expressions presented in this chapter are programmed to determine the condensation heat transfer coefficient and the applications of the program are illustrated with representative examples.

8-1 HEAT TRANSFER COEFFICIENTS FOR CONDENSATION OF VAPORS

We present below the recommended correlations for the calculation of the average heat transfer coefficient for the filmwise condensation of saturated, stationary (i.e., low velocity) vapors which are free of noncondensable gas on a cold surface maintained at a uniform temperature lower than the saturation temperature of the vapor. Condensation on following geometries is considered.

Vertical plate or vertical tube

Single horizontal tube

Horizontal tube banks

Depending on the rate of flow of condensate, the flow regime may be laminar or turbulent, which affect the heat transfer coefficient. Both of these cases are examined.

Vertical Plate or Vertical Tube

Consider condensation of saturated vapor over a vertical plate or vertical tube of height L, maintained at a uniform temperature T_w which is lower than the saturation temperature T_v of the vapor. The average heat transfer coefficient h_m for the laminar and turbulent flow of condensate are determined from the following expressions.

Laminar flow (Re ≤ 1800).

The average heat transfer coefficient h_m is calculated from the Nusselt's equation with a modified coefficient.

$$h_m = 1.13\left(\frac{g\,\rho_l^2\,h_{fg}\,k_l^3}{\mu_l(T_v - T_w)L}\right)^{1/4} \tag{8-1}$$

where

g = 9.81 m/s^2 is gravitational acceleration

T_w = Wall temperature

T_v = Saturation temperature

L = Height

k_l = Liquid thermal conductivity

μ_l = Liquid viscosity

ρ_l = Liquid density

h_{fg} = Latent heat of condensation

The physical properties of the condensate is evaluated at the film temperature taken as $(T_w + T_v)/2$ and it is assumed that $\rho_v << \rho_l$.

The condensate Reynolds number is given by

$$Re = \frac{4L\, h_m(T_v - T_w)}{\mu_l\, h_{fg}} \tag{8-2}$$

Turbulent flow (Re > 1800).

For turbulent film condensation the average heat transfer coefficient is determined from the empirical Kirkbride (1934) formula

$$h_m \left(\frac{\mu_l^2}{k_l^3\, \rho_l^2\, g} \right)^{1/3} = 0.0077\, Re^{0.4} \tag{8-3}$$

Reynolds number given by Eq. (8-2) is introduced into Eq. (8-3) and the resulting expression is solved for h_m to give

$$h_m = \left\{ 0.0077 \left(\frac{k_l^3\, \rho_l^2\, g}{\mu_l^2} \right)^{1/3} \left(\frac{4L(T_f - T_w)}{\mu\, h_{fg}} \right)^{2/5} \right\}^{5/3} \tag{8-4}$$

The heat transfer coefficients given by Eqs. (7-1) and (7-4) are used in the present software.

Single Horizontal Tube

The average heat transfer coefficient h_m for laminar film condensation on a horizontal tube is given by the Nusselt formula

$$h_m = 0.725 \left(\frac{g\, \rho_l^2\, h_{fg}\, k_l^3}{\mu_l(T_v - T_w)D} \right)^{1/4} \tag{8-5}$$

where D is the outside diameter of the tube and it is assumed that $\rho_v \ll \rho_l$. For condensation on a single horizontal tube the condensate flow is laminar for all practical surfaces. Properties are evaluated at the film temperature.

Horizontal Tube Banks

Condenser design generally involves horizontal tubes arranged in vertical tiers in such a way that the condensate from one tube drains onto the tube below. If it is assumed that the drainage from one tube flows smoothly onto the tube below, then for a N number of vertical tiers each having tubes of diameter D, the average heat transfer coefficient h_m is given by

$$h_m = 0.725\left(\frac{g\,\rho_l^2\,h_{fg}\,k_l^3}{\mu_l(T_v - T_w)D}\right)^{1/4}\frac{1}{N^{1/4}} \tag{8-6}$$

Properties are evaluated at the film temperature. This relation generally yields a conservative result since some turbulence and disturbance of condensate film are unavoidable during drainage which increase the heat transfer coefficient.

8-2 COMPUTER SOLUTIONS

The analytic expressions given previously for the determination of film condensation heat transfer coefficient are used in the program FILMWISE CONDENSATION for the calculation of film condensation heat transfer coefficients. The physical properties needed for the computations are considered readily obtainable from the property tables available in the standard heat transfer books. Therefore, the property data used in the following examples are not repeated in the statement of the problem, instead they are shown in the Tables of results. Graphical support is also available to plot the heat transfer coefficient as a function of the temperature difference between the vapor and metal surface temperatures. The applications of this program are illustrated with representative examples.

Example 8-1. Air free saturated low velocity steam at temperature T_v=90C (i.e., at pressure 70.14 kPa) is condensed on the outer surface of a 1.5 m-long, 2.5 cm-OD vertical tube maintained at a uniform temperature T_w=70C. Figure below shows the geometry. Calculate the average heat transfer coefficient h_m over the entire length of the vertical tube and using the graphical support examine the variation of h_m with the temperature difference.

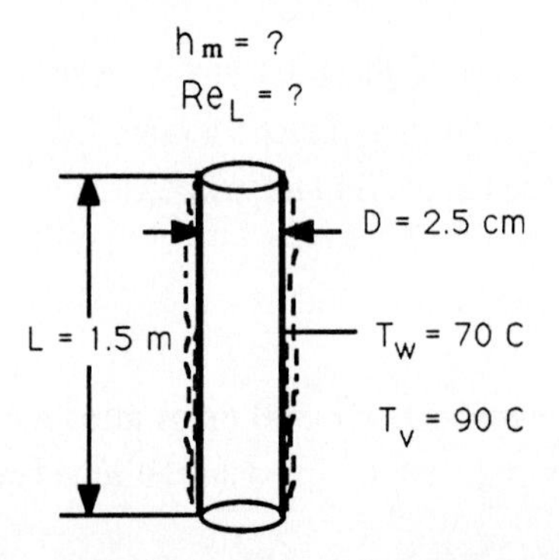

Solution. Table below shows the computer INPUT and OUTPUT data for this example after the program has been run.

FILMWISE	CONDENSATION	
Wall temperature	: C	70.000
Vapor temperature	: C	90.000
Geometry	: Vertical Plate	
Height	: m	1.500
Liquid thermal conductivity	: W/m•K	0.668
Liquid viscosity	: kg/m•s	0.000335000
Liquid density	: kg/m^3	974.000
Latent heat of condensation	: kJ/kg	2309.000
Heat transfer coefficient	= W/m²•K	5677.694
Reynolds number	=	880.815

In the example, we have h_m=5677.694 W/m^2·K and Re_L=880.815. The condensate flow is laminar since Re_L<1800.

The following figure shows the graphical representation of the heat transfer coefficient plotted against the height, which is the characteristic length of the condensing surface.

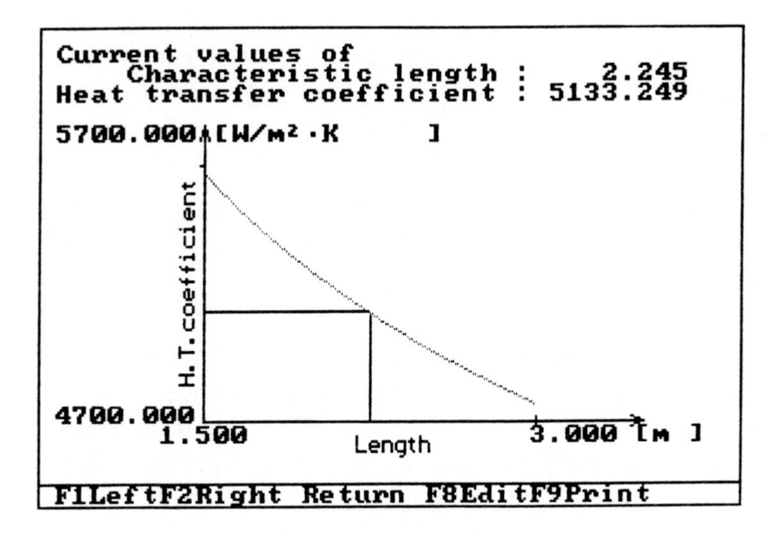

Finally, the following Table, obtained from the above graphical representation, gives the values of the heat transfer coefficient for several different characteristic lengths.

FILMWISE CONDENSATION

Geometry : Vertical Plate

Temp.difference [m]	H.T. coeficient [W/m²•K]
1.500	5677.694
1.700	5501.700
1.900	5349.992
2.100	5217.134
2.200	5156.530
2.400	5045.092
2.600	4944.742
2.800	4853.639
2.990	4774.353

Example 8-2. Air-free saturated steam at T_V=50°C (P=12.35 kPa) condenses on the outside surface of a 2.5 cm-OD, 2-m-long vertical tube maintained at a uniform temperature T_W=30°C by the flow of cooling water through the tube. Assuming film condensation, calculate the average condensation heat transfer coefficient over the entire length of the tube. Figure below shows the geometry.

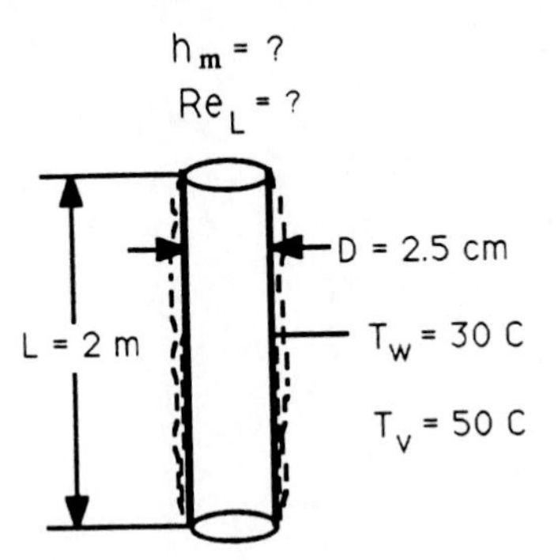

Solution. Table below shows the computer INPUT and OUTPUT data for this example after the program has been run. The solution for the problem gives the average condensation heat transfer coefficient over the entire length of the tube as h_m=4354.641 W/m^2·K and the condensate Reynolds number at the bottom of the tube is Re_L=441.931. The condensate flow is laminar because Re_L<1800.

```
                FILMWISE    CONDENSATION
          Wall temperature : C                     30.000
         Vapor temperature : C                     50.000
                  Geometry : Vertical Plate
                    Height : m                      2.000

Liquid thermal conductivity : W/m•K                 0.628
          Liquid viscosity : kg/m•s                 0.000655000
            Liquid density : kg/m^3               994.000
Latent heat of condensation : kJ/kg              2407.000

   Heat transfer coefficient = W/m²•K             4354.641
             Raynolds number =                     441.931
```

Example 8-3. Saturated air-free steam at T_V=75°C (P≅38.58 kPa) condenses on a 0.5 m-by-0.5 m vertical plate maintained at a uniform temperature T_W=45°C. Calculate the average film condensation heat transfer coefficient h_m over the entire length of the plate, Figure below shows the geometry.

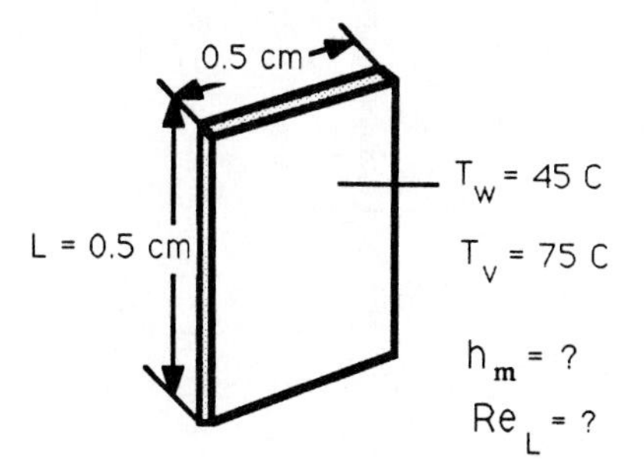

Solution. Table below shows the computer INPUT and OUTPUT data for this example.

```
                FILMWISE    CONDENSATION
          Wall temperature : C                     45.000
         Vapor temperature : C                     75.000
                  Geometry : Vertical Plate
                    Height : m                      0.500

Liquid thermal conductivity : W/m•K                 0.651
          Liquid viscosity : kg/m•s                 0.000471000
            Liquid density : kg/m^3               985.500
Latent heat of condensation : kJ/kg              2358.500

   Heat transfer coefficient = W/m²•K             6150.220
             Raynolds number =                     332.189
```

The problem similar to the previous example except the geometry is a vertical plate and the temperatures are different. The entering of the input data is straightforward. The solution for the problem gives the average condensation heat transfer coefficient as h_m=6150.22 W/m^2·K and the condensate Reynolds number at the bottom of the plate as Re=332.189.

Example 8-4. Saturated ammonia vapor at T_v=−5°C (P=0.3528 Mpa) condenses on the outer surface of a 0.75-m-long, 1.27-cm-OD vertical tube maintained at a uniform temperature T_w=-15°C. Calculate the average condensation heat transfer coefficient h_m over the entire length of the tube. Figure below shows the geometry.

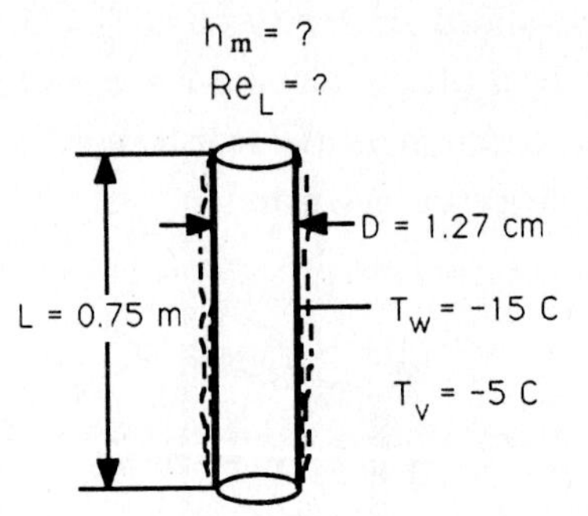

Solution. Table below shows the computer INPUT and OUTPUT data for this example.

```
                 FILMWISE    CONDENSATION
        Wall temperature  : C                         -15.000
       Vapor temperature  : C                          -5.000
                Geometry  : Vertical Tube
                  Height  : m                           0.750

Liquid thermal conductivity  : W/m•K                    0.543
          Liquid viscosity  : kg/m•s                    0.000247000
            Liquid density  : kg/m^3                  653.600
Latent heat of condensation  : kJ/kg                 1280.000

   Heat transfer coefficient  = W/m²•K               5243.419
            Raynolds number  =                        497.541
```

The condensate flow for this example is also laminar.

Example 8-5. Air-free saturated steam at T_v=90°C (P=70.14 kPa) condenses on the outer surface of a 2.5-cm-OD, 6-m-long vertical tube maintained at a uniform temperature T_w=30°C. Calculate the average heat transfer coefficient over the entire length of the tube. Figure below shows the geometry.

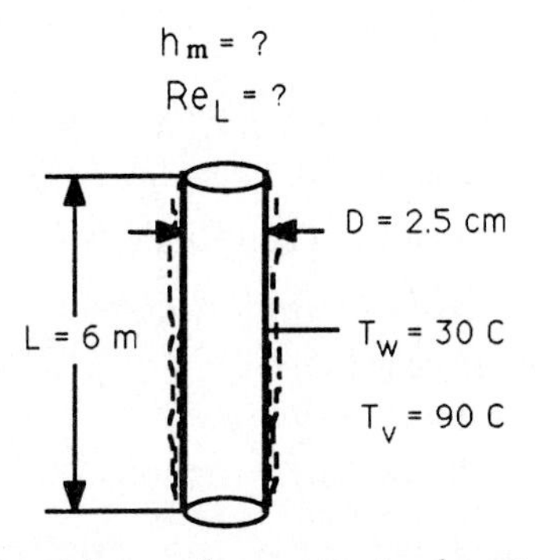

Solution. Table below shows the computer input and output data for this example after the program has been run.

FILMWISE	CONDENSATION	
Wall temperature :	C	30.000
Vapor temperature :	C	90.000
Geometry :	Vertical Tube	
Height :	m	6.000
Liquid thermal conductivity :	W/m•K	0.651
Liquid viscosity :	kg/m•s	0.000471000
Liquid density :	kg/m^3	985.000
Latent heat of condensation :	kJ/kg	2358.000
Heat transfer coefficient =	W/m²•K	6540.269
Raynolds number =		8479.952

Condensate flow is turbulent because $Re_L > 1800$.

Example 8-6. Air-free saturated steam at T_V=60°C (P=19.94 kPa) condenses on the outer surface of 100 horizontal tubes with 2.5-cm OD and 2 m long, arranged in a

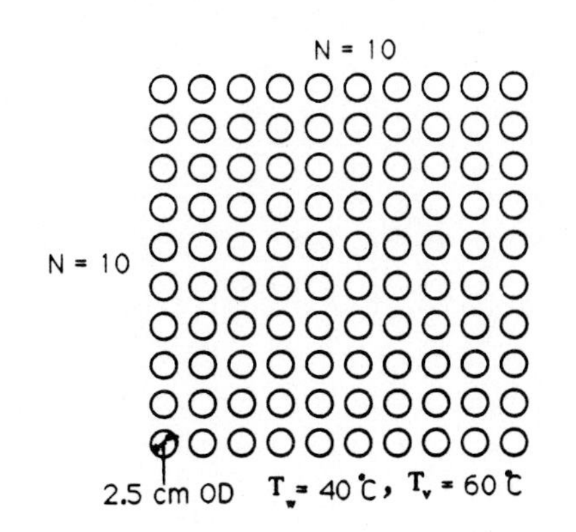

10 by 10 square array. The surface of the tubes is maintained at a uniform temperature T_W=40°C. Calculate the average condensation heat transfer coefficient for the entire tube bundle, the total rate of heat transfer. Figure above shows the geometry.

Solution. Table below shows the computer input and output data for this example after the program has been run.

```
                  FILMWISE    CONDENSATION
          Wall temperature : C                        40.000
         Vapor temperature : C                        60.000
                  Geometry : Horizontal Tube Banks
     Tube outside diameter : cm                        2.500
   Number of vertical tiers:                          10
Liquid thermal conductivity : W/m•K                    0.640
           Liquid viscosity : kg/m•s                   0.000563000
             Liquid density : kg/m^3                 990.000
Latent heat of condensation : kJ/kg                 2383.000

  Heat transfer coefficient = W/m²•K                4927.416
```

Chapter
NINE
NUCLEATE BOILING

When a surface is in contact with a liquid and maintained at a temperature above the saturation temperature of the liquid, boiling may occur. If the heated surface is immersed in a quiescent body of liquid and below the free surface, the boiling process is called pool boiling. Figure 9-1 shows a plot of the surface heat flux against the temperature difference between the heater surface and the liquid saturation temperature for pool boiling. Note that, there are three distinct regimes of boiling, which include free_convection, nucleate boiling and film boiling regimes. For most practical applications, the nucleate boiling regime is most desirable because large heat fluxes are obtainable with small temperature differences.An examination of this figure reveals that, in the nucleate boiling regime, the heat flux increases rapidly with increasing temperature difference until the peak heat flux or the so called burn-out heat flux is reached. As soon as the peak heat

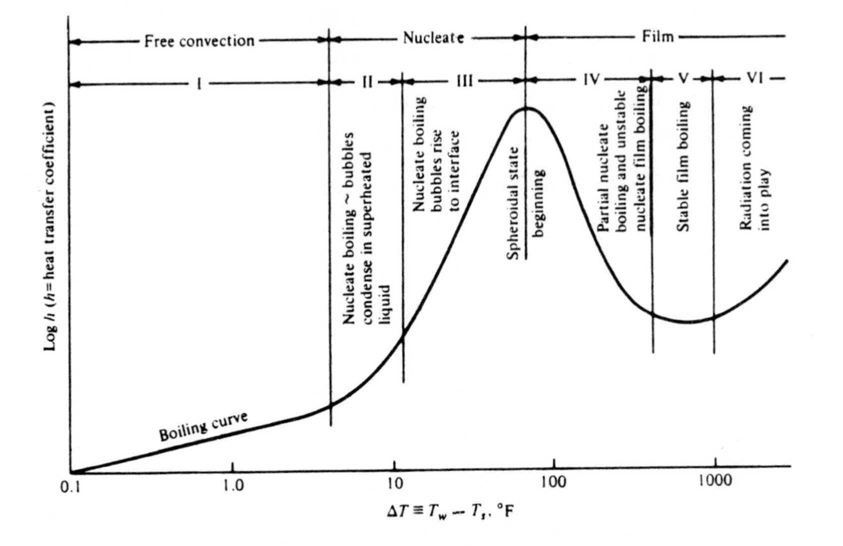

Figure 9-1 Principal boiling regimes in pool boiling of water at atmospheric pressure and saturation temperature T_s from an electrically heated platinum wire. (From Farber and Scorah (1984)).

flux is exceeded, extremely high temperature differences are needed to realize the resulting heat flux, because the boiling regime changes into the film boiling.

In this chapter, recommended correlations are presented for determining the nucleate boiling heat flux and the peak heat flux. The film boiling will be the subject of the next chapter.

The correlations presented in this chapter are programmed for calculating the nucleate boiling heat flux and the peak heat flux. The application of the computer program is illustrated with representative examples.

9-1 NUCLEATE BOILING HEAT FLUX

The heat flux q over the entire nucleate boiling regime can be calculated from the empirical correlation developed by Rohsenov (1952)

$$\frac{C_{pl}\,\Delta T}{h_{fg}\,Pr_l^n} = C_{sf}\left[\frac{q}{\mu_l h_{fg}}\sqrt{\frac{g_c\,\sigma^*}{g(\rho_l - \rho_v)}}\right]^{1/3} \tag{9-1}$$

where

c_{pl} = specific heat of saturated liquid, J/(kg · °C) or Btu/(lb · °F)

C_{sf} = constant to be determined from experimental data depending on heating surface-fluid combination

h_{fg} = latent heat of vaporization, J/kg or Btu/lb

g = gravitational acceleration m/s^2 or ft/h^2

g_c = gravitational acceleration conversion factor, 1 kg · m/(N · s^2) or lb · ft/(lbf · h^2); it is not needed in the SI system.

Pr_l = $c_{pl}\mu_l/k_l$ = Prandtl number of saturated liquid

q = boiling heat flux, W/m^2 or Btu/(h · ft^2)

ΔT = $T_w - T_s$, temperature difference between wall and saturation temperature, °C or °F

μ_l = viscosity of saturated liquid, kg/(m · s) or lb/(ft · h)

ρ_l, ρ_v = density of liquid and saturated vapor, respectively, kg/m^3 or lb/ft^3

σ^* = surface tension of liquid-vapor interface, N/m or lbf/ft

In Eq. (9-1) the exponent n and the coefficient C_{sf} are the two provisions to adjust the correlation for the liquid-surface combination. Table 9-1 lists the experimentally determined values C_{sf} for a variety of liquid-surface combinations. The value of n should be taken as 1 for water and 1.7 for all the other liquids shown in Table 9-1.

Table 9-1 Values of the coefficient C_{sf} of Eq. (9-1) for various liquid-surface combinations

Liquid-surface combination	C_{sf}
Water-copper	0.0130
Water-scored copper	0.0068
Water-emery-polished copper	0.0128
Water-emery-polished, paraffin-treated copper	0.0147
Water-chemically etched stainless steel	0.0133
Water-mechanically polished stainless steel	0.0132
Water-ground and polished stainless steel	0.0080
Water-Teflon pitted stainless steel	0.0058
Water-platinum	0.0130
Water-brass	0.0060
Benzene-chromium	0.0100
Ethyl alcohol-chromium	0.0027
Carbon tetrachloride-copper	0.0130
Carbon tetrachloride-emery-polished copper	0.0070
n-Pentane-emery-polished copper	0.0154
n-Pentane-emery-polished nickel	0.0127
n-Pentane-emery-rubber copper	0.0074
n-Pentane-lapped copper	0.0049

Table 9-2 gives the values of the vapor-liquid surface tension for a variety of liquids.

For a given surface heat flux q, the temperature difference $\Delta T \equiv T_w - T_s$ can readily be calculated from Eq. (9-1). However, if the temperature difference is given and the corresponding surface heat flux is required, it is desirable to rearrange equation (9-1). Therefore this equation is rewritten and rearranged in the following forms:

$$\Delta T = B(Aq)^{1/3} \tag{9-2}$$

or

$$q = \frac{1}{A}\left(\frac{\Delta T}{B}\right)^3 \tag{9-3}$$

where

$$A = \frac{h_{fg}\, Pr_l^n}{C_{pl}}\, C_{sf} \tag{9-4}$$

Table 9-2 Values of liquid-vapor surface tension σ^* for various liquids

Liquid	Saturation temperature		Surface tension	
	°F	°C	$\sigma^* \times 10^4$ lbf/ft	$\sigma^* \times 10^3$ N/m
Water	32	0.0	51.8	75.6
Water	60	15.56	50.2	73.2
Water	100	37.78	47.8	69.7
Water	200	93.34	41.2	60.1
Water	212	100	40.3	58.8
Water	320	160	31.6	46.1
Water	440	226.7	21.9	31.9
Water	560	293.3	11.1	16.2
Water	680	360	1.0	1.46
Water	705.4	374.11	0.0	0
Sodium	1618	881.1	77	11.2
Potassium	1400	760	43	62.7
Rubidium	1270	687.8	30	43.8
Cesium	1260	682.2	20	29.2
Mercury	675	357.2	27	39.4
Benzene (C_6H_6)	176	80	19	27.7
Ethyl alcohol (C_2H_6O)	173	78.3	15	21.9
Freon 11	112	44.4	5.8	8.5

$$B = \frac{1}{\mu_l h_{fg}} \sqrt{\frac{\sigma^*}{g(\rho_l - \rho_v)}} \tag{9-5}$$

Equations (9-1) and (9-2) are programmed to calculate the temperature difference ΔT when the boiling heat flux q is given and to calculate the boiling heat flux when the temperature difference is given, respectively.

9-2 PEAK HEAT FLUX

The correlation given by Eq. (9-3) gives the heat flux in nucleate boiling, but it can not predict the peak heat flux, which is of interest because of burnout considerations.

For example, if the applied surface heat flux exceeds the peak heat flux, the boiling regime changes from the nucleate boiling to the stable film boiling. As apparent from Fig. 9-1, temperature difference may increase by at least an order of magnitude and the resulting surface temperature may exceed the melting point of the heating surface.

The following expression developed by Ded and Leinhard (1972), Sun and Leinhard (1970), Leinhard and Dhir (1973) can be used to predict the peak heat flux, q_{max}:

$$q_{max} = F(L') \times 0.131 \rho_v^{1/2} h_{fg}[\sigma^* g(\rho_l - \rho_v)]^{1/4} \tag{9-6}$$

where the various quantities are defined previously except F(L'), which is a correction factor for the effects of heater geometry and size. The factor F(L') depends on the dimensionless characteristic length L' of the heater, defined as

$$L' = L\sqrt{\frac{g(\rho_l - \rho_v)}{\sigma^*}} \tag{9-7}$$

where L is the characteristic dimension of the heater and the other quantities are as defined previously. Table 9-3 lists the recommended values of the factor F(L') for geometries such as an infinite plate, horizontal cylinder, and a sphere and defines the characteristic length dimension for the geometries considered.

Table 9-3 Correction factor F(L') for use in Eq. (9-6)

Heater geometry	F(L')	Remarks
1. Infinite flat plate facing up	1.14	$L' \geq 2.7$; L is the heater width or diameter
2. Horizontal cylinder	$0.89 + 2.27e^{-3.44\sqrt{L'}}$	$L' \geq 0.15$; L is the cylinder radius
3. Large sphere	0.84	$L' \geq 4.26$; L is the sphere radius
4. Small sphere	$\frac{1.734}{(L')^{1/2}}$	$0.15 \leq L' \leq 4.26$; L is the sphere radius
5. Large finite body	~0.90	$L' \geq 4$; $L = \frac{\text{volume}}{\text{surface area}}$

In the present program Eq. (9-6) is used to predict the peak heat flux, q_{max}.

9-3 COMPUTER SOLUTIONS

The empirical expressions given by Eqs. (9-2) and (9-3) are programmed to calculate the temperature difference ΔT and the nucleate boiling heat flux, q, respectively. Equations (9-6) and (9-7) are used to compute the peak heat flux, q_{max}.

The correction factors F(L') listed in Table 9-6 are built into the program.

The liquid-surface coefficients C_{sf} should be obtained from Table 9-1 for the given liquid-surface combination and entered into the program as an INPUT data. Other physical properties needed for the computations are obtainable from the property tables available in the standard heat transfer books.

The program NUCLEATE BOILING can handle boiling heat transfer calculations for boiling on following geometries:

Large infinite body

Flat square plate facing up

Horizontal cylinder, and

Sphere.

In addition, for each of these geometries, the following two different types of calculations can be performed:

(i) For a specified temperature difference ΔT and liquid-surface combination, the computer calculates the corresponding heat flux q and the peak heat flux q_{max}. If the selected ΔT belongs to the nucleate boiling regime, q should be small than q_{max}; otherwise the selected ΔT is beyond that for the nucleate boiling regime

(ii) For a specified surface heat flux q and liquid-surface combination, the computer calculates the corresponding ΔT and the peak heat flux q_{max}. Again, q should be smaller than q_{max}.

In addition graphical support is available to plot the surface heat flux for nucleate boiling against the temperature difference between the surface and vapor saturation temperatures.

The following examples illustrate the application of the program *NUCLEATE BOILING*.

Example 9-1. Saturated water at T_s=100C is boiled on a flat square copper pan of sides L=10 cm. The surface of the heated pan is at a uniform temperature T_w=109C. Calculate the surface heat flux, the peak heat flux and use the graphical support to plot the boiling heat flux against the temperature difference between the pan surface and water saturation temperatures.

Solution. Table below shows the computer INPUT and OUTPUT data for this example after the program has been run.

```
                         NUCLEATE  BOILING
                 Geometry :  Flat square plate facing up
    Heater width or diameter :  cm                    10.000
 Latent heat of vaporization :  kJ/kg               2257.000
     Specific heat of liquid :  J/kg•K              4216.000
         Viscosity of liquid :  kg/m•s                 0.0002820
           Density of liquid :  kg/m^3               960.000
            Density of vapor :  kg/m^3                 0.600
             Surface tension :  N/m                    0.0588000
                Coefficients   Csf:  0.01300  n:  1.000
    Prandtl number of liquid :                         1.740
      Temperature difference :  C                      9.000

           Surface heat flux =  W/m²              104540.175
              Peak heat flux =  W/m²              1.2663E+06
```

The main features of entering the INPUT data and the use of graphical support is now described.

Entering the INPUT data. The first INPUT data in this table is for setting up the type of geometry. Move the cursor next to the *Geometry* and press the SPACEBAR. Each time the SPACEBAR is pressed the geometry changes, successively, to *Flat square plate facing up*, *Horizontal cylinder*, *Sphere* and *Large finite body.* Press the ENTER key when desired geometry is displayed. For this example the geometry is *Flat square plate facing up.*

The next item is the heater width which is entered as 10 cm.

The following six input data are for the physical properties of liquid and vapor, which are obtainable from the standard property data.

The coefficient C_{sf} for water-copper combination is obtained from Table 9-1 as C_{sf}=0.013 and the exponent n for water is taken as n=1 as specified previously. The value of the Prandtl number for water is then entered.

The next item following the Prandtl number of liquid can alternate between *Temperature difference* and the *Boiling heat flux* (i.e., surface heat flux) when the cursor is moved over this quantity and the SPACEBAR is pressed. In this example we have chosen it as *Temperature difference,* because the temperature difference between the surface and vapor temperatures (i.e., $T_w - T_v = 109 - 100 = 9$) is known. (In Example 9-3 we consider the case of prescribed boiling heat flux).

The last two items are the computer OUTPUT, which give the *Surface heat flux* (i.e., boiling heat flux) and the *Peak heat flux.* The peak heat flux should be higher than the surface heat flux. Otherwise, the selected temperature difference is beyond the range of nucleate boiling regime.

Graphical Support. The graphical support is available to plot the surface heat flux against the temperature difference $T_W - T_V = \Delta T$ as shown in figure below.

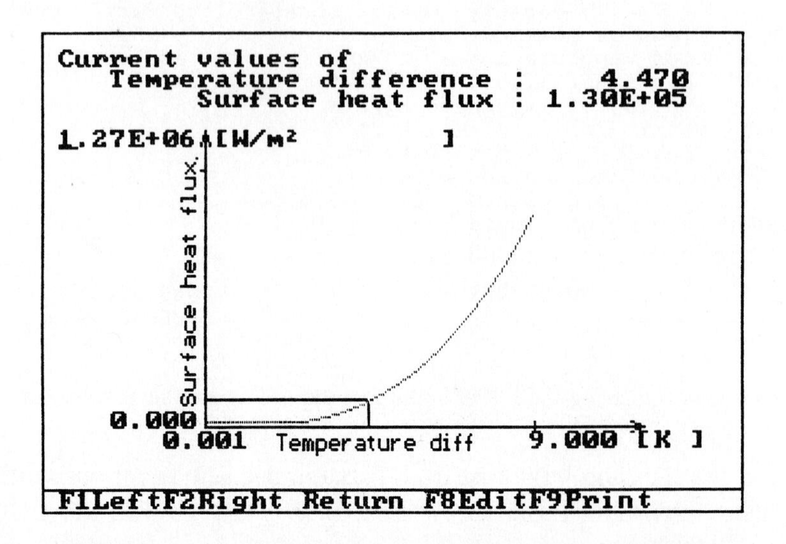

Example 9-2. Repeat example 9-1 for the case of a brass heating pan.

Solution. The input data for this problem are the same as that for the Example 9-1, except for the brass surface the coefficient C_{sf} is changed from 0.013 to 0.006. The resulting computer output given in Table below shows that the surface heat flux has changed only very little.

```
                           NUCLEATE  BOILING
                Geometry :  Flat square plate facing up
   Heater width or diameter :  cm                     10.000
 Latent heat of vaporization :  kJ/kg                2257.000
     Specific heat of liquid :  J/kg•K               4216.000
         Viscosity of liquid :  kg/m•s                  0.0002820
           Density of liquid :  kg/m^3                960.000
            Density of vapor :  kg/m^3                  0.600
             Surface tension :  N/m                     0.0588000
                Coefficients   Csf:  0.00600   n:  1.000
     Prandtl number of liquid :                         1.740
      Temperature difference :  C                       9.000

           Surface heat flux =  W/m²               1.0633E+06
              Peak heat flux =  W/m²               1.2663E+06
```

Example 9-3. Saturated water at a temperature T_S=100°C is boiled with a copper heating element of diameter D=1 cm. If the surface heat flux of the element is 400 kW/m^2, calculate the surface temperature. What is the maximum heat flux?

Solution. This problem is different from the two previous problems in that the *Surface heat flux* (i.e., Boiling Heat Flux is given, the *Temperature difference* is unknown. Table below shows the INPUT and OUTPUT data for this problem after the program has been run.

```
                       NUCLEATE  BOILING
             Geometry :  Flat square plate facing up
 Heater width or diameter :  cm                     1.000
Latent heat of vaporization :  kJ/kg               2257.000
   Specific heat of liquid :  J/kg•K               4216.000
       Viscosity of liquid :  kg/m•s                 0.0002820
         Density of liquid :  kg/m^3                960.000
          Density of vapor :  kg/m^3                  0.600
           Surface tension :  N/m                     0.0588000
              Coefficients   Csf:  0.01300  n:  1.000
    Prandtl number of liquid :                        1.740
         Boiling heat flux :  kW/m²                 400.000

   Temperature Difference =  C                       14.077
          Peak heat flux =  W/m²                 1.2663E+06
```

The INPUT data for the first 10 items from the top of Table above including the Prandtl number, are entered as discussed previously.

Move the cursor over the item following the Prandtl number and press the SPACEBAR. Each time the SPACEBAR is pressed, it changes alternately between *Temperature difference* and *Boiling heat flux*. In this example the boiling heat flux is specified as 400 kW/m^2. Therefore, press the ENTER key when *Boiling heat flux* is displayed and then enter its value 400 kW/m^2.

The last two items are the computer OUTPUTS. Note that, in this case the *Temperature difference* and the *Peak heat flux* are the computer OUTPUT.

Example 9-4. Water at atmospheric pressure and saturation temperature is boiled by using an electrically heated, copper circular disc of diameter D=20 cm with heated surface facing up. The surface of the heating element is maintained at a uniform temperature T_W=110°C. Calculate the surface heat flux and the peak heat flux.

Solution. Table below shows the computer INPUT and OUTPUT data for this problem after the program has been run. All the INPUT data are obtained and entered as discussed previously. In this problem the temperature difference, $T_W - T_S = 110 - 100 = 10$C is specified. Therefore, the item following the Prandtl number for liquid is set to *Temperature difference*. Then, the last two items, *Surface heat flux* and *Peak heat flux* are the OUTPUT data.

```
                              NUCLEATE  BOILING
              Geometry :  Flat square plate facing up
  Heater width or diameter :  cm                   20.000
 Latent heat of vaporization :  kJ/kg              2257.000
    Specific heat of liquid :  J/kg•K             4216.000
       Viscosity of liquid :  kg/m•s                0.0002820
         Density of liquid :  kg/m^3              960.000
          Density of vapor :  kg/m^3                0.600
           Surface tension :  N/m                   0.0588000
              Coefficients    Csf:  0.01300   n:  1.000
   Prandtl number of liquid :                       1.740
     Temperature difference :  C                   10.000

         Surface heat flux =  W/m²             143402.160
            Peak heat flux =  W/m²             1.2663E+06
```

Example 9-5. An electrically heated, copper, spherical heating element of diameter D=20 cm is immersed in water at atmospheric pressure and temperature. The surface element is maintained at a uniform temperature T_W–115°C. Calculate the surface heat flux and the peak heat flux.

Solution. In this example the temperature difference $T_W - T_V$=115-100=15C is specified and the surface heat flux (i.e., Boiling heat flux) is to be determined. The INPUT data are obtained and entered as discussed previously. The item following the Prandtl number is set to *Temperature difference* and its value 15C is entered. The last two items, *Surface heat flux* and *Peak heat flux* are the OUTPUTS. Table below gives the INPUT and OUTPUT data for this example.

```
                              NUCLEATE  BOILING
              Geometry :  Sphere
          Sphere radius :  cm                   10.000
 Latent heat of vaporization :  kJ/kg              2257.000
    Specific heat of liquid :  J/kg•K             4216.000
       Viscosity of liquid :  kg/m•s                0.0002820
         Density of liquid :  kg/m^3              960.000
          Density of vapor :  kg/m^3                0.600
           Surface tension :  N/m                   0.0588000
              Coefficients    Csf:  0.01300   n:  1.000
   Prandtl number of liquid :                       1.740
     Temperature difference :  C                   15.000

         Surface heat flux =  W/m²             483982.290
            Peak heat flux =  W/m²             933080.853
```

Example 9-6. Repeat Example 9-5 for a horizontal cylindrical element of diameter 20 cm.

Solution. Table below shows the INPUT and OUTPUT data for this example after the program has been run.

```
                      NUCLEATE  BOILING
                 Geometry :  Horizontal cylinder
          Cylinder radius :  cm                      10.000
Latent heat of vaporization :  kJ/kg                 2257.000
    Specific heat of liquid :  J/kg•K                4216.000
        Viscosity of liquid :  kg/m•s                   0.0002820
          Density of liquid :  kg/m^3                 960.000
           Density of vapor :  kg/m^3                   0.600
            Surface tension :  N/m                      0.0588000
               Coefficients   Csf:  0.01300  n:   1.000
   Prandtl number of liquid :                           1.740
     Temperature difference :  C                       15.000

          Surface heat flux =  W/m²               483982.290
             Peak heat flux =  W/m²               988621.381
```

The problem is similar to that given by Example 9-5 except the heating element is a horizontal cylinder of diameter 20 cm instead of a sphere. Note that the surface heat flux is similar to that of Example 9-5, but the peak heat flux is slightly higher.

Chapter

TEN

FILM BOILING

As illustrated in Fig. 9-1 the nucleate boiling region ends and the unstable film boiling region begins after the peak heat flux is reached. No analysis is available for the prediction of heat flux as a function of the temperature difference $T_W - T_S$ in this unstable region until the minimum point in the boiling curve is reached and the stable film boiling region starts. In stable film boiling regions V and VI, the heating surface is separated from the liquid by a vapor layer across which heat must be transferred. Since the thermal conductivity of the vapor is low, large temperature differences are needed for heat transfer in this region. The stable film boiling regime has numerous applications, among others in the boiling of cryogenic fluids. In this chapter we present the expressions for the determination of the film boiling heat transfer coefficient. The correlations presented here are then used in the program FILM BOILING to calculate the film boiling heat transfer coefficient.

10-1 FILM BOILING HEAT TRANSFER COEFFICIENT

Bromley (1950) developed analytic expression for the determination of heat transfer coefficient for stable film boiling on the outside of a horizontal cylinder. The resulting expression for the average heat transfer coefficient h_o for stable film boiling on the outside of a horizontal cylinder, in the absence of radiation, is given by

$$h_o = 0.62\left[\frac{k_v^3\,\rho_v\,(\rho_l - \rho_v)\,g\,h_{fg}}{\mu_v\,d_o\,\Delta T}\left(1 + \frac{0.4C_{pv}\,\Delta T}{h_{fg}}\right)\right]^{1/4} \tag{10-1}$$

where

h_o = average heat transfer coefficient in the absence of radiation, W/m^2·C

C_{pv} = specific heat of saturated vapor, J/kg·C

d_o = outside diameter of tube, m

g = gravitational acceleration, m/s^2

h_{fg} = latent heat of evaporization, J/kg

k_v = thermal conductivity of vapor, W/(m·°C)

ΔT = $T_W - T_S$ = temperature difference, °C

T_W = wall surface temperature, °C

T_s = saturation temperature of liquid, °C

ρ_l = density of liquid

ρ_v = density of vapor at saturation temperature

μ_v = viscosity of vapor

With film boiling, when surface temperature is sufficiently high, the radiation effects become important. For such cases the radiation heat transfer coefficient h_r can be estimated approximately from the following formula

$$h_r = \frac{1}{1/\varepsilon + 1/\alpha - 1} \frac{\sigma(T_W^4 - T_S^4)}{T_W - T_S} \tag{10-2}$$

where α = absorptivity of liquid

ε = emissivity of hot tube

σ = Stefan-Boltzmann constant

T_W = wall temperature

T_S = saturation temperature of liquid

when h_r is smaller than h_o, the following expression can be used to determine the total heat transfer coefficient h_m which includes the effects of radiation

$$h_m = h_o + \frac{3}{4} h_r \tag{10-3}$$

10-2 COMPUTER SOLUTIONS

The expressions given above are used in the program FILM BOILING to calculate the film boiling heat transfer coefficient. In the following examples we illustrate the application of this program for the calculation of film boiling heat transfer coefficient. Graphical support is also available to examine the variation of film boiling heat transfer coefficient with the surface temperature of the heating element.

Example 10-1. Water at saturation temperature and atmospheric pressure is boiled with an electrically heated platinum wire of diameter D_o=0.2 cm in the stable film boiling regime with a temperature difference $T_W - T_S$ = 654°C. Calculate the film boiling heat transfer coefficient h_m. Absorptivity of water can be taken as α=1 and the emissivity of the wire as ε=1.

Solution. Table below shows the computer INPUT and OUTPUT data after the program has been run.

FILM BOILING

Wall temperature	:	C	754.000
Saturation temperature	:	C	100.000
Latent heat of vaporization	:	kJ/kg	2257.000
Specific heat of vapor	:	J/kg•K	2085.000
Viscosity of vapor	:	kg/m•s	0.0000243
Density of liquid	:	kg/m^3	960.600
Density of vapor	:	kg/m^3	0.314
Thermal conductivity of vapor	:	W/m•K	0.050500
Tube diameter	:	cm	0.200
Absorptivity of liquid	:		1.000
Emissivity of Tube	:		1.000
Heat transfer coefficient	=	W/m²•K	336.537

The first INPUT data is the heater wall temperature which is taken as

$$T_W = 654 + 100 = 754C$$

since the saturation temperature of water is 100C.

This program can also be used to determine the film boiling heat transfer coefficient h_o in the absence of radiation if we set very small values for the absorptivity and emissivity, say α=0.001 and ε=0.001. Table below shows the computer INPUT and OUTPUT for the same problem with very small values of emissivity and absorptivity.

FILM BOILING

Wall temperature	:	C	754.000
Saturation temperature	:	C	100.000
Latent heat of vaporization	:	kJ/kg	2257.000
Specific heat of vapor	:	J/kg•K	2085.000
Viscosity of vapor	:	kg/m•s	0.0000243
Density of liquid	:	kg/m^3	960.600
Density of vapor	:	kg/m^3	0.314
Thermal conductivity of vapor	:	W/m•K	0.050500
Tube diameter	:	cm	0.200
Absorptivity of liquid	:		0.001
Emissivity of Tube	:		0.001
Heat transfer coefficient	=	W/m²•K	265.460

Note that setting a very small value either for ε or α gives essentially the same result.

The graphical support is also available to examine the effects of diameter on the film boiling heat transfer coefficient. We consider the case with negligible radiation effects (i.e., ε=α=0.001). Press the F8 function key to activate the graphical support.

The screen displays the coordinate axes, "Heat Transfer Coefficient, $W/m^2 \cdot K$" versus "Length (i.e., diameter), cm" and fixes some ranges of the heat transfer coefficient and the length. These ranges of the *heat transfer coefficient* and the *length,* automatically selected by the computer, can be changed by moving the cursor next to the quantity to be altered, typing the new value and then pressing the enter key. To plot the results, press the F8 function key once more. The following figure shows a typical graph.

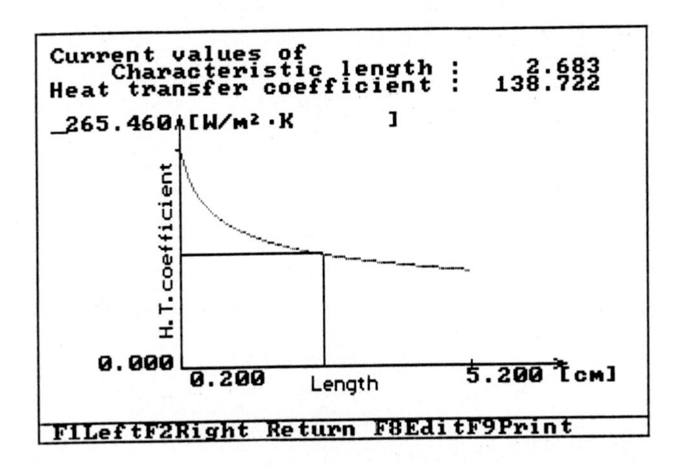

By moving the cursor over the graph, the data at the cursor location can be stored in the computer memory each time the RETURN key is pressed. By pressing the F9 print key, a hard copy of such data can be retrieved as shown in the following table.

FILM BOILING

Length [cm]	H.t.coefficient [W/m²•K]
0.200	265.460
0.502	210.908
1.005	177.298
1.509	160.192
2.012	149.070
2.515	140.979
3.019	134.696

Example 10-2. Repeat example (9-1) for the case of a heating element of diameter D=0.6 cm.

Solution. The problem is exactly the same as that given in Example 9-1 except in the present case the heater diameter should be changed to D=0.6 cm. Table below showing the computer INPUT and OUTPUT data for this problem gives the film boiling heat transfer coefficient including the effects of radiation

FILM BOILING

Wall temperature :	C	754.000
Saturation temperature :	C	100.000
Latent heat of vaporization :	kJ/kg	2257.000
Specific heat of vapor :	J/kg•K	2085.000
Viscosity of vapor :	kg/m•s	0.0000243
Density of liquid :	kg/m^3	960.600
Density of vapor :	kg/m^3	0.314
Thermal conductivity of vapor :	W/m•K	0.050500
Tube diameter :	cm	0.600
Absorptivity of liquid :		1.000
Emissivity of Tube :		1.000
Heat transfer coefficient =	W/m²•K	272.792

we note that the heat transfer coefficient is reduced from h_m=336.537 W/m^2·K to 272.792 W/m^2·K by increasing the radius of the heating element from 0.2 cm to 0.6 cm.

Table below show the solution of this problem by neglecting the radiation effects, that is by taking very small emissivity and absorptivity (i.e., $\varepsilon=\alpha=0.001$ is chosen).

FILM BOILING

Wall temperature :	C	754.000
Saturation temperature :	C	100.000
Latent heat of vaporization :	kJ/kg	2257.000
Specific heat of vapor :	J/kg•K	2085.000
Viscosity of vapor :	kg/m•s	0.0000243
Density of liquid :	kg/m^3	960.600
Density of vapor :	kg/m^3	0.314
Thermal conductivity of vapor :	W/m•K	0.050500
Tube diameter :	cm	0.600
Absorptivity of liquid :		0.001
Emissivity of Tube :		0.001
Heat transfer coefficient =	W/m²•K	201.714

Example 10-3. Water at saturation temperature and atmospheric pressure is boiled with an electrically heated, horizontal platinum wire of diameter D=0.2 cm. Boiling takes place with a temperature difference of $T_W - T_S$ = 454°C in the stable film boiling regime. Calculate the film boiling heat transfer coefficient and the heat flux. Assume $\varepsilon=1$, $\alpha=1$.

Solution. Next two tables show the solution of this problem with and without the radiation effects, respectively.

```
                                 FILM  BOILING
                  Wall temperature :  C                       554.000
            Saturation temperature :  C                       100.000
       Latent heat of vaporization :  kJ/kg                  2257.000
            Specific heat of vapor :  J/kg•K                 2026.000
                Viscosity of vapor :  kg/m•s                    0.0000206
                 Density of liquid :  kg/m^3                  960.600
                  Density of vapor :  kg/m^3                    0.365
    Thermal conductivity of vapor :  W/m•K                     0.044200
                     Tube diameter :  cm                        0.200
            Absorptivity of liquid :                            1.000
                Emissivity of Tube :                            1.000

        Heat transfer coefficient =  W/m²•K                   322.141
```

```
                                 FILM  BOILING
                  Wall temperature :  C                       554.000
            Saturation temperature :  C                       100.000
       Latent heat of vaporization :  kJ/kg                  2257.000
            Specific heat of vapor :  J/kg•K                 2026.000
                Viscosity of vapor :  kg/m•s                    0.0000206
                 Density of liquid :  kg/m^3                  960.600
                  Density of vapor :  kg/m^3                    0.365
    Thermal conductivity of vapor :  W/m•K                     0.044200
                     Tube diameter :  cm                        0.200
            Absorptivity of liquid :                            0.001
                Emissivity of Tube :                            0.001

        Heat transfer coefficient =  W/m²•K                   280.135
```

For the later case, very small values are chosen for ε and α (i.e., $\alpha=\varepsilon=0.001$) in order to eliminate the radiation effects for all practical purposes.

Chapter

ELEVEN

BLACKBODY RADIATION

A body at any temperature above absolute zero emits thermal radiation in all wavelengths in all possible directions into space. The concepts of blackbody is an idealized situation that serves to compare the emission and absorption characteristics of real bodies.

A blackbody is considered to absorb all incident radiation from all directions at all wavelengths without reflecting, transmitting, or scattering it. The radiation emission by a blackbody at any temperature T is the maximum possible emission at that temperature.

The term BLACK should be distinguished from its common usage regarding the blackness of a surface to visual observations. The human eye can detect blackness only in the visible range of the spectrum. For example, an object such as ice is bright to the eye but is almost black for long-wave thermal radiation. However, a blackbody is perfectly black to thermal radiation for all wavelengths from $\lambda=0$ to $\lambda=\infty$.

In this chapter we present the expression for determining the blackbody spectral emissive power as a function of wavelength and emissivity, and the equation presented here is used in the program *BLACKBODY RADIATION FUNCTION.*

11-1 SPECTRAL BLACKBODY EMISSIVE POWER

The spectral blackbody radiation emitted per unit surface area in all directions into the hemispherical space is called the SPECTRAL BLACKBODY EMISSIVE POWER. It represents the radiation energy emitted by a blackbody at an absolute temperature T per unit area per unit time per unit wavelength about λ in all directions into the hemispherical space, and is determined from the Planck [1959] blackbody radiation law as

$$E_{b\lambda}(T) = \frac{c_1}{\lambda^5\{\exp[c_2/(\lambda T)] - 1\}} \quad W/(m^2 \cdot \mu m) \qquad (11\text{-}1)$$

where $c_1 = 2\pi hc^2 = 3.743 \times 10^8$ W·μm^4/m^2

$c_2 = \frac{hc}{k} = 1.4387 \times 10^4$ μm·K

T = absolute temperature, K

λ = wavelength, μm.

Thus, given the temperature T and the wavelength λ, the spectral blackbody emissive power $E_{b\lambda}(T)$ can be calculated.

11-2 COMPUTER SOLUTIONS

Equation (11-1) is programmed to calculate $E_{b\lambda}(T)$ for a given temperature T over a specified wavelength band and present the results in the graphical form as a function of the wavelength. When the program *BLACKBODY RADIATION FUNCTION* is activated the screen displays the coordinate system *Emissive flux* vs *Wavelength* as shown in the figure below.

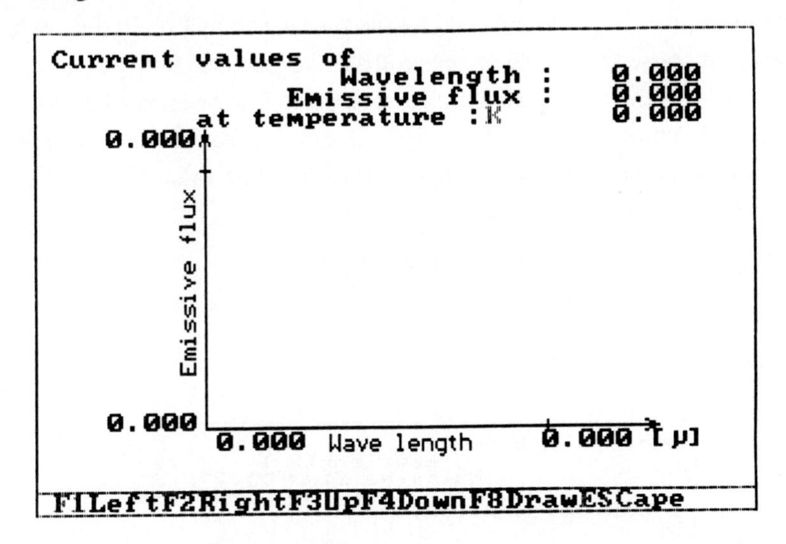

Note that, each time the SPACEBAR is pressed, the unit of temperature successively changes to C, F, K and so on. Therefore, first, the unit of temperature is selected and the ENTER key is pressed. Then the numerical value of temperature is entered.

Next, the beginning and end values of the *Emissive flux* and *Wavelength* intervals are selected by moving the cursor to the appropriate locations and entering the desired values. When all INPUT data are completed, the F8 function key is pressed in order to plot the emissive flux as a function of the wavelength.

To present the results in the graphical form, the computer divides the specified wavelength band into 150 equal parts and calculates $E_{b\lambda}(T)$ at each of these subdivisions. If the cursor is moved along the graph, the current values of λ and $E_{b\lambda}(T)$ automatically replace the old ones. In the graphical form, each time the ENTER key is pressed, the exact values of λ and $E_{b\lambda}(T)$ corresponding to the cursor location are stored in the memory of the computer. Up to 150 such values can be stored. A hard copy of λ and $E_{b\lambda}(T)$ stored in the memory can be obtained by pressing the F9 function key.

We now illustrate the application of the program *BLACKBODY RADIATION FUNCTION* with representative examples.

Example 11-1. Plot the spectral blackbody emissive power $E_{b\lambda}(T)$ at temperature T=1000K over the wavelength band from λ=1 μm to λ=20 μm. Determine the maximum value of $E_{b\lambda}(T)$ and the corresponding wavelength.

Solution. When the coordinate system *Emissive flux* vs *Wavelength* appears, the cursor is over the unit K. Pres the ENTER key to set Kelvin as temperature unit and then type the numerical value of temperature 1000. Suppose we select the intervals for the Emissive flux as 0 to 1.5×10^4 and for the wavelength as 1 to 20 μm. When the F8 function key is pressed the results are presented in the graphical form and the exact values of the wavelength and the emissive flux corresponding to the cursor location appear on the upper right corner of the figure. We move the cursor along the curve until the maximum value of the Emissive flux and the corresponding value of wavelength appears on the screen. For this example we have

$$\lambda = 2.913 \text{ μm}$$

$$E_{b\lambda}(T) = 12900 \text{ W/m}^2\cdot\text{μm}$$

as shown in the figure below.

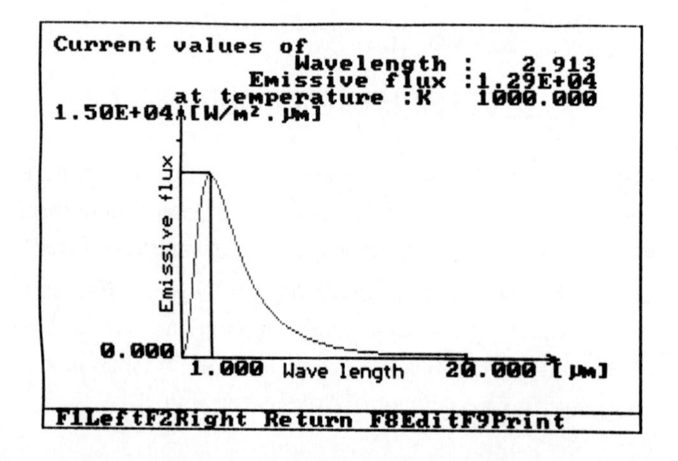

Since this result is based on the fact that the wavelength interval 0-20 μm is divided into 150 equal subintervals, each time the SPACEBAR is pressed the cursor moves $\frac{20}{150}$ μm. Therefore, the accuracy of the result can be improved by narrowing down the intervals of wavelength and emissive power and repeating the calculations. The following figure shows a similar plot but for the beginning and end values for the emissive power selected as $10^4 - 1.3 \times 10^4$ and for the wavelength 2 – 4 μm.

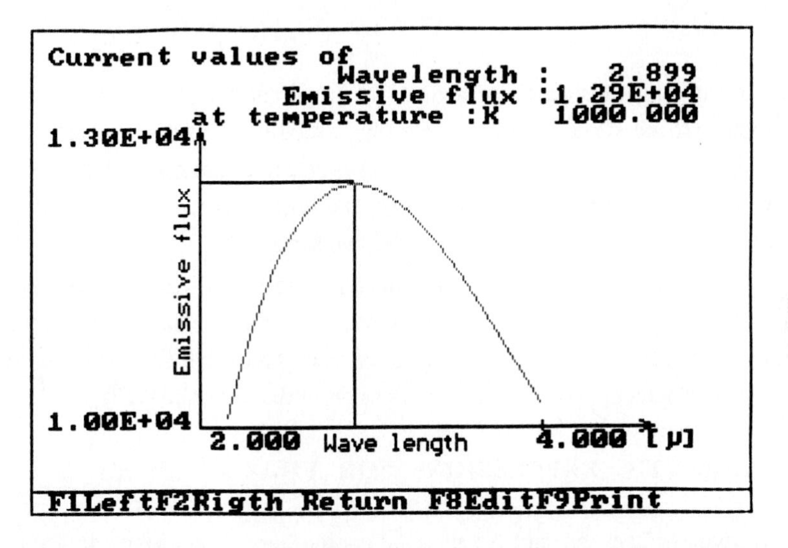

In this case, by moving the cursor along the graph and observing the value of the emissive flux appearing at the upper right corner of the graph, the maximum value of the emissive flux and the corresponding wavelength are determined as

$$\lambda = 2.899 \ \mu m$$

$$E_{b\lambda}(T) = 12900$$

The above zooming procedure can be repeated by narrowing down the ranges of the emissive power and the wavelength and more refined results obtained.

Chapter

TWELVE

VIEW FACTORS

Radiation exchange between surfaces is of practical interest in many engineering applications. If the surfaces are separated by a medium that does not absorb, emit, or scatter radiation, the orientation of surfaces plays an important role in radiation exchange, because radiation propagates in straight lines, and if one surface cannot see another, there is no direct exchange of radiation between the two surfaces.

A quantity of practical interest that plays a significant role in governing the radiation heat exchange between surfaces is the View Factor. In this chapter we present the analytic expressions for the evaluation of view factor between surfaces for simple geometric arrangements and use these expression in the program *VIEW FACTORS*.

12-1 ANALYTIC EXPRESSIONS FOR VIEW FACTORS

Consider two surfaces A_1 and A_2 viewing each other. The surface area A_1 will emit radiation in all directions in the hemispherical space and only a certain fraction of this emitted energy will strike surface A_2 directly. The View Factor $F_{1,2}$ represents the percentage of the total radiation energy that leaves surface A_1 in all direction into the hemisphere that strikes the surface A_1 directly. A comprehensive complication of view factors are given by Siegel and Howell (1981), Hamilton and Morgan (1952), Mackey, et al. (1943) and Howell (1982). We present below the view factors for some simple geometric arrangements.

Two equal parallel opposed rectangles

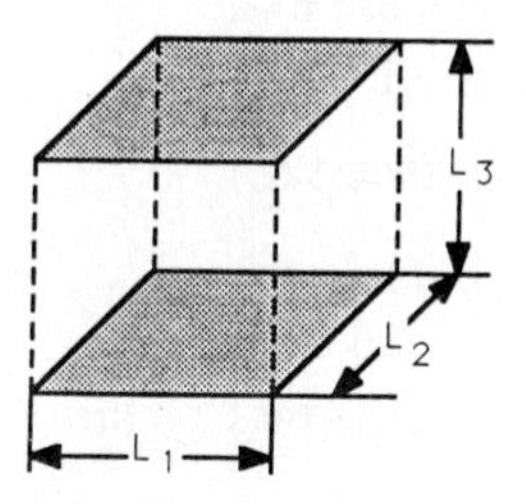

Figure 12-1 Two equal parallel opposed rectangles

Input data: Length = L_1, Width = L_2, Distance between plates = L_3.

By using the dimensionless parameters

$$X = \frac{L_1}{L_3}, \qquad Y = \frac{L_2}{L_3}$$

the analytical expressions for the view factor $F_{1,2}$ are

For $X \to \infty$, $Y \to \infty$: $$F_{1,2} = 1 \tag{12-1a}$$

For $X \to \infty$: $$F_{1,2} = \sqrt{1 + \left(\frac{1}{Y}\right)^2} - \frac{1}{Y} \tag{12-1b}$$

For $Y \to \infty$: $$F_{1,2} = \sqrt{1 + \left(\frac{1}{X}\right)^2} - \frac{1}{Y} \tag{12-1c}$$

For X and Y both finite:

$$F_{1,2} = \frac{2}{\pi YX}\left\{ \ln \sqrt{\frac{(1 + X^2)(1 + Y^2)}{1 + X^2 + Y^2}} + X\sqrt{1 + Y^2}\tan^{-1}\frac{X}{\sqrt{1 + Y^2}} - X\tan^{-1}X + Y\sqrt{1 + X^2}\tan^{-1}\frac{Y}{\sqrt{1 + X^2}} - Y\tan^{-1}Y \right\} \tag{12-1d}$$

Two coaxial parallel circular discs

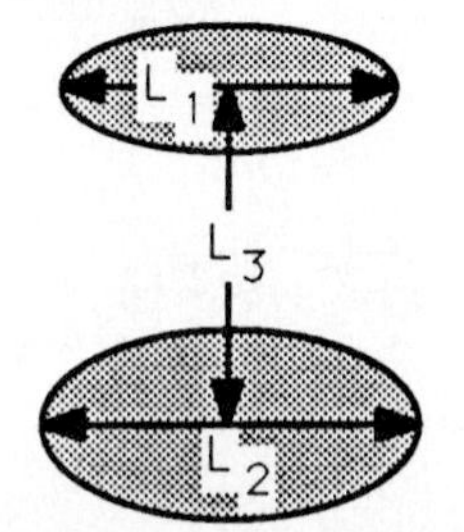

Figure 12-2 Two coasial parallel circular discs

Input data: Diameter of the first disc = L_1

Diameter of the second disc = L_2

Distance between disc = L_3

By introducing the dimensionless parameters

$$R_1 = \frac{L_2}{L_1}, \qquad R_2 = \frac{L_3}{L_1}$$

the analytical expressions for the view factor $F_{1,2}$ are

$$F = \left(1 + R_1^2 + R_2^2 - \sqrt{(1 + R_1^2 + R_2^2)^2 - 4R_1^2R_2^2}\right)\Big/2 \tag{12-2a}$$

$$F_{1,2} = F/R_1^2, \qquad F_{2,1} = F/R_2^2 \tag{12-2b}$$

Two perpendicular rectangles with a common edge

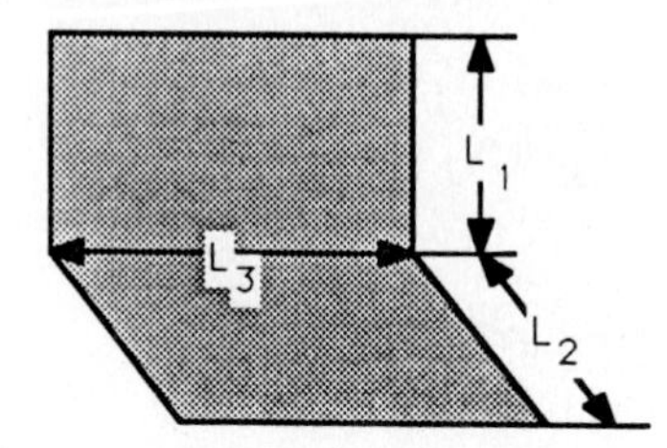

Figure 12-3 Two perpendicular rectangles with a common edge

Input data: Height of the first rectangle = L_1

Height of the second rectangle = L_2

Length of the common side = L_3

The following dimensionless parameters are defined

$$H = \frac{L_1}{L_3}, \qquad W = \frac{L_2}{L_3}$$

Then the view factor are

$$F = \frac{1}{\pi W}\left(W \tan^{-1}\frac{1}{W} + H \tan^{-1}\frac{1}{H} - \sqrt{H^2 + W^2}\tan^{-1}\frac{1}{\sqrt{H^2 + W^2}}\right.$$

$$+\frac{1}{4}\ln\left\{\frac{(1+W^2)(1+H^2)}{1+W^2+H^2}\left[\frac{W^2(1+W^2+H^2)}{(1+W^2)(W^2+H^2)}\right]^{W^2}\left[\frac{H^2(1+H^2+W^2)}{(1+H^2)(H^2+W^2)}\right]^{H^2}\right\}\Bigg) \quad (12\text{-}3a)$$

$$F_{1,2}=\frac{F}{W} \quad (12\text{-}3b)$$

$$F_{2,1}=\frac{F}{H} \quad (12\text{-}3c)$$

12-2 COMPUTER SOLUTIONS

The expressions for the view factors given above are used in the program VIEW FACTORS. In this section we illustrate the application of this program with representative examples.

Example 12-1. Two aligned, parallel, square plates 0.6 m by 0.6 m are separated by a distance L=0.3 m as illustrated in the figure below. Calculate the view factor $F_{1\text{-}2}$ and $F_{2\text{-}1}$ between them.

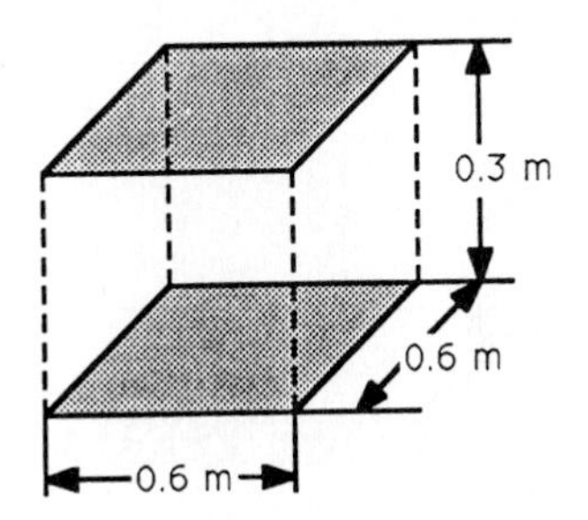

Solution. Table below shows the computer INPUT and OUTPUT data after the program has been run.

```
                         VIEW FACTORS

              Geometry :  Equal parallel opposed rectangles

      Length of the rectangle :  m               0.600
       Width of the rectangle :  m               0.600
Distance between the rectangles :  m             0.300

         View factor (1-2) =                      0.415
         View factor (2-1) =                      0.415
```

The first INPUT data is the setting up the type of geometry. Each time the press bar is pressed the screen displays, successively, the following three geometric arrangements: *Equal parallel opposed rectangles, Perpendicular rectangles with a common edge, Coaxial parallel discs.*

The length, width and distance are introduced in a similar manner as described in previous chapters.

Figure below shows the graphical support for this program, giving a plot of F_{1-2} plotted against the distance between the plates over the range of this distance from 0 to 0.3 m.

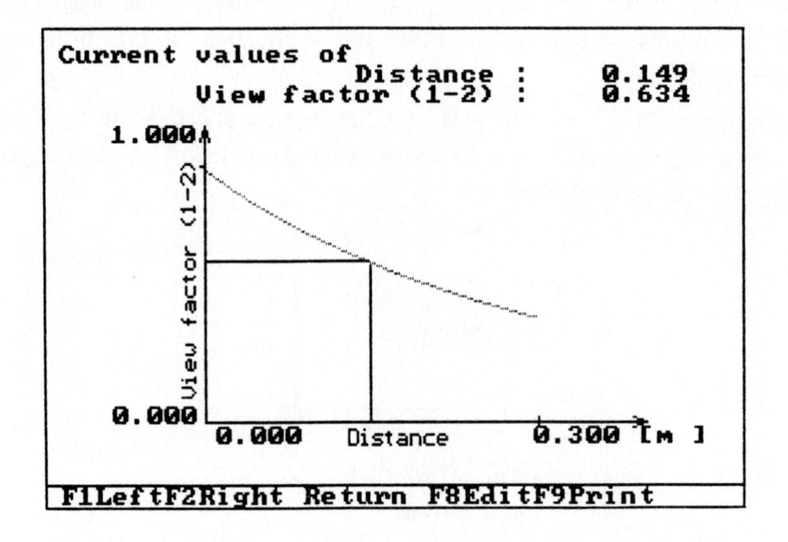

To exit from the graphics mode, first press F8 and then the ESC keys.

Example 12-2. Two parallel circular discs of equal diameter d=0.5 m separated by a distance L=0.25 m as illustrated in figure below. Calculate the view factor F_{1-2}.

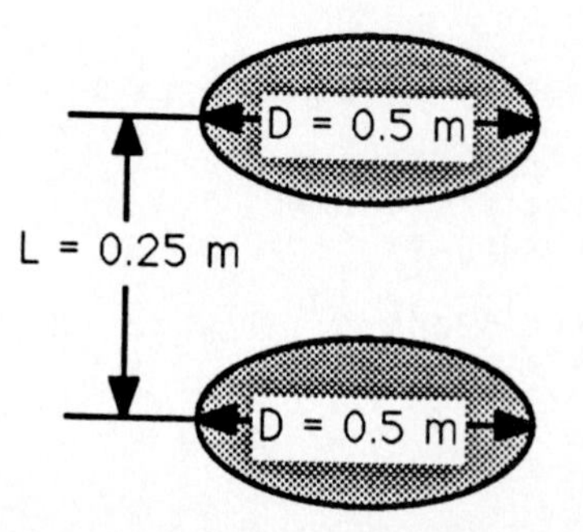

Solution. Table below shows the INPUT and OUTPUT data after the program has been run.

```
                    VIEW FACTORS

          Geometry :  Coaxial parallel disks

   Diameter of the first disk :  m              0.500
  Diameter of the second disk :  m              0.500
   Distance between the disks :  m              0.250

           View factor (1-2) =                  0.610
           View factor (2-1) =                  0.610
```

Example 12-3. Determine the view factors F_{1-2} and F_{2-1} between two rectangular surfaces A_1 and A_2 for the geometric arrangement shown in figure below.

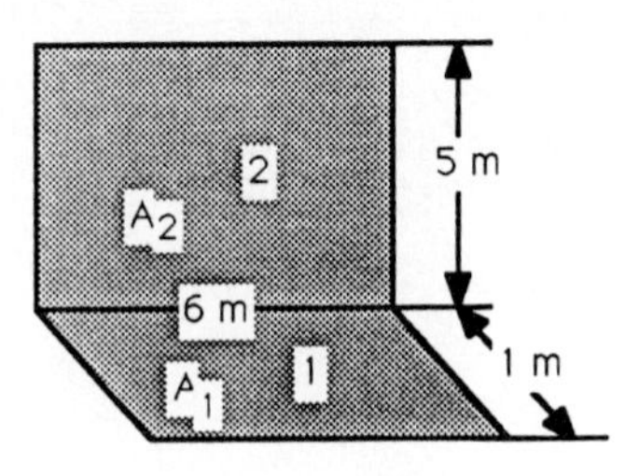

Solution. Table below shows the computer INPUT and OUTPUT data for this example.

```
                         VIEW FACTORS

             Geometry :  Perpend. rectangles with a common edge

  Height of the first rectangle :  m              1.000
 Height of the second rectangle :  m              5.000
      Length of the common side :  m              6.000

              View factor (1-2) =                 0.397
              View factor (2-1) =                 0.079
```

Example 12-4. Two aligned, parallel, square plates 0.6 m by 0.6 m are separated by a distance L=0.3 m as illustrated in figure below. Calculate the view factors F_{1-2} and F_{2-1}.

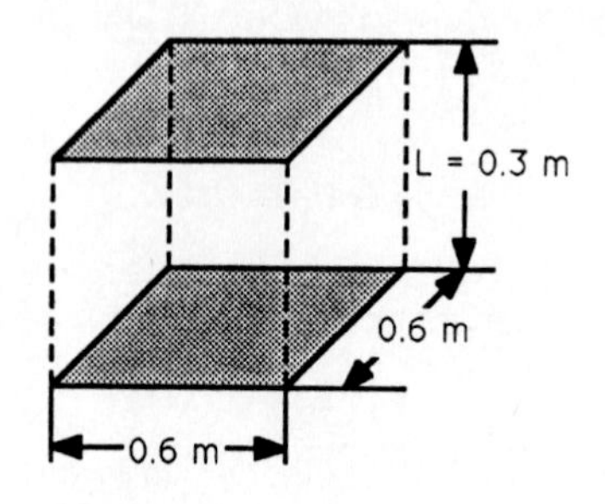

Solution. Table below shows the computer INPUT and OUTPUT data for this example.

```
                              VIEW FACTORS

                     Geometry :  Equal parallel opposed rectangles

         Length of the rectangle :  m                         0.600
          Width of the rectangle :  m                         0.600
 Distance between the rectangles :  m                         0.300

               View factor (1-2) =                            0.415
               View factor (2-1) =                            0.415
```

Chapter

THIRTEEN

RADIATION SHIELDS

The radiation heat transfer between two surfaces can be reduced significantly if a radiation shield made of low-emissivity material is placed between them. In this chapter we examine the radiation heat transfer between two parallel plates, long coaxial cylinders and concentric spheres and the resulting reduction in the heat transfer rate when one or more layers of radiation shields are placed between them. The analytical expressions for determining the total heat flow rate are given and the use of the program *RADIATION SHIELDS* for calculating the radiation heat transfer is illustrated with examples.

13-1 ANALYTIC EXPRESSIONS FOR HEAT FLOW THROUGH RADIATION SHIELDS

To illustrate this matter we first consider radiation heat transfer between two parallel plates, long coaxial cylinders and concentric spheres.

Let A_1 and A_2 be the surface areas, T_1 and T_2 the absolute temperatures and ε_1 and ε_2 be the emissivities of the inner and outer surfaces, respectively. We assume the surfaces are all opaque. Figure 13-1 shows the geometry.

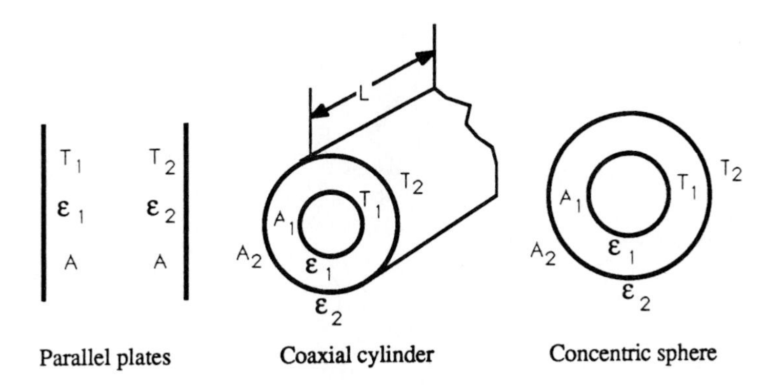

Figure 13-1 Nomenclature for radiation exchange between two surfaces

The system consisting of n radiation shields of surface areas $A_1, A_2, ..., A_n$ placed between the *first surface* of area A_o and the *last surface* of area A_{n+1} as shown in Figure 13-2

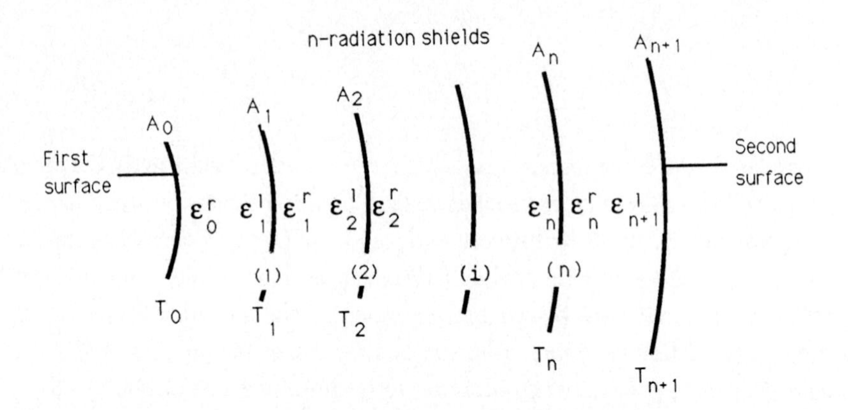

Figure 13-2 Nomenclature for radiation heat transfer through a system with n-shields

Radiation heat transfer Q through the system is given by

$$Q = C_s \frac{\left(\frac{T_o}{100}\right)^4 - \left(\frac{T_{n+1}}{100}\right)^4}{R}, \quad W \tag{13-1}$$

where

$$C_s = 5.67, \quad W/m^2 K^4$$

and R is the total resistance

$$R = \sum_{i=1}^{n} R_i \tag{13-2}$$

The individual resistances R_i given by

$$R_i = \frac{1}{A_{i-1}\,\varepsilon_{i-1}^r} + \frac{1}{A_i}\left(\frac{1}{\varepsilon_i^l} - 1\right) \qquad i = 1,2,...,n \tag{13-3}$$

The surface areas A_i (i=1 or 2) can be written as

$$A_i = \pi\, L\, D_i \qquad i = 1,2 \tag{13-4a}$$

for a cylinder, and

$$A_i = \pi D_i^2 \qquad i = 1,2 \qquad (13\text{-}4b)$$

for a sphere, where

L = length of the cylinder

D_i = diameter of surface i

For convenience in the nomenclature we used the superscripts l and r over the emissivity to denote "left" and "right" respectively. For example, ε_2^l refers to the emissivity of the left surface of A_2 with reference to the illustration shown in Fig. 13-2.

Once the total heat flow rate Q is known, the temperature T_k of any shield layer, k, is determined from the expression

$$T_k = 100\left[\left(\frac{T_o}{100}\right)^4 - \frac{Q\,R_k}{C_s}\right] \qquad (13\text{-}5)$$

where R_k is the resistance to radiation associated with the shields no. 1 through no. k, and is given by

$$R_k = \sum_{i=1}^{k} R_i$$

The above equations are the basis for calculating the total heat flow rate and the shield temperatures in a system consisting of n-radiation shields between the inner and outer surfaces.

13-2 COMPUTER SOLUTIONS

The above equations are used in the program RADIATION SHIELDS to calculate the total heat flow rate Q and the shield temperatures in a system consisting of up to 8 shields between two parallel plates, coaxial cylinders and concentric spheres. The application is now illustrated with representative examples.

Example 13-1. A radiation shield is placed between two large parallel plates as illustrated in the figure below.

The *first plate* has an emissivity 0.8 and temperature 1000K.

The *second plate* has an emissivity 0.5 and temperature 500K.

The *radiation shield* has an emissivity 0.05 at both surfaces.

Determine the radiation heat transfer per unit area through the system and the shield temperature.

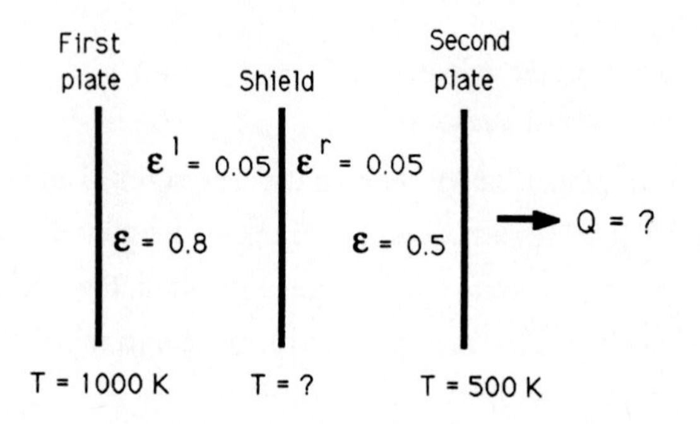

Solution. Table below shows the computer INPUT and OUTPUT data for this example after the program has been run.

```
                    RADIATION SHIELDS
                    Geometry : Parallel plates
                        Area : m²                              1.000
       First plate temperature : K                          1000.000
        First plate emissivity :                               0.800

             Number of shields : 1

Emissivity of     Emissivity of
First surface     Second surface                       Temperatures =
  0.050             0.050                              K      857.142


        Last plate temperature : K                           500.000
         Last plate emissivity :                               0.500

               Heat flow rate = W                           1288.636
```

The first INPUT data is the setting up of the type of geometry by pressing the SPACEBAR. Each time the SPACEBAR is pressed, the geometry changes successively, to *Parallel plates, Long cylinders* and *Concentric spheres.* Press the ENTER key when *Parallel plates* is displayed.

The next item is the AREA. Set the UNIT for the area key pressing the SPACEBAR when the cursor is next to the area. Each time the SPACEBAR is pressed the unit for the area changes, successively, to m^2, cm^2, ft^2 and in^2. Press the RETURN key when m^2 appears, since the area is given in meter square. Then enter the value of the area as 1, press the SPACEBAR and then the RETURN key.

The next two items are the INPUT data for the first plate; they include the temperature 1000K and emissivity 0.8, which are entered in the usual manner.

The number of shields is specified as 1 and the emissivity of the first surface (i.e., left surface) 0.05 and of the second surface (i.e., right surface) 0.05 are entered. The *Temperature* of the shield followed by the equal sign, is an unknown and will be given as the OUTPUT after the program has been run. However, the unit of the shield temperature can be selected by pressing the SPACEBAR. each time the SPACEBAR is pressed, the unit changes, successively, to K, C and F. Press ENTER key when K is displayed.

The next two items are the INPUT data associated with the second plate. Its temperature 1000K and emissivity 0.5 are entered.

The last item the *Heat flow rate* followed by the equal sign is an unknown and will be given after the program has been run.

Graphical Support. The graphical support is available to examine the effects of the *first* or the *last* plate temperature on the heat flow rate across the system. For example, figure below shows the variation of the heat flow rate with the variation of the *first* plate temperature from 500.0K to 1000.0K.

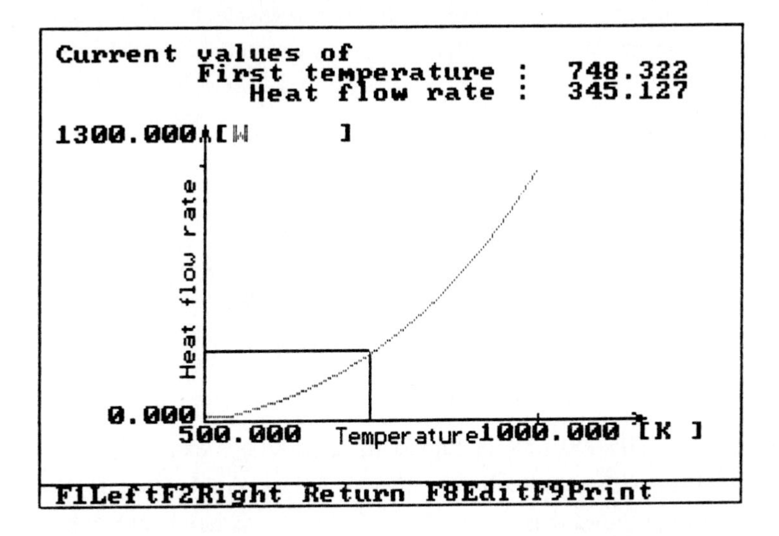

To obtain graphical support for the *Last* plate temperature, press F8 function key to change to the *Edit mode*, move the cursor over the word *First* in this graph press the SPACEBAR. The word *First temperature* will change to *Last temperature.* Press F8 function key, the following graph will be obtained for the *Last plate temperature.*

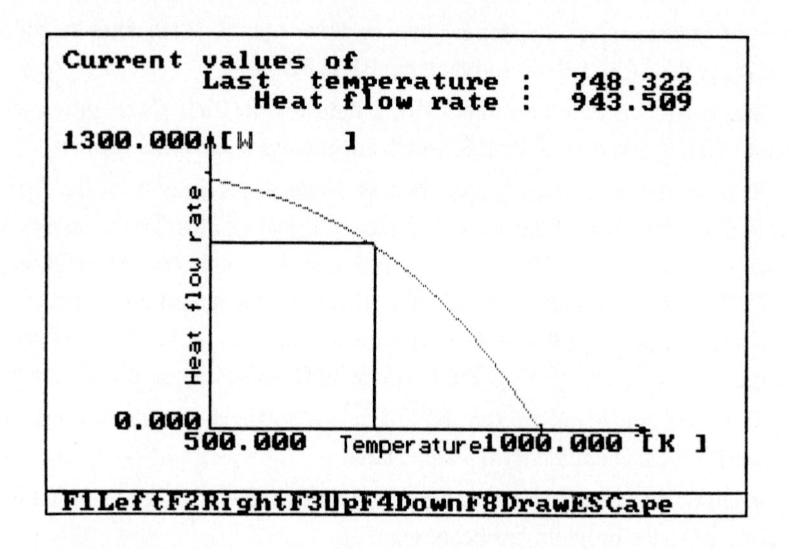

Example 13-2. Consider a radiation shield between two large parallel plates as illustrated in figure below

The *first plate* is at 800K and has an emissivity 0.9.

The *second plate* is at 300K and has an emissivity 0.5.

A radiation shield of emissivity 0.1 at both surfaces is placed between them. Calculate the heat transfer rate per unit area through the system without and with the shield.

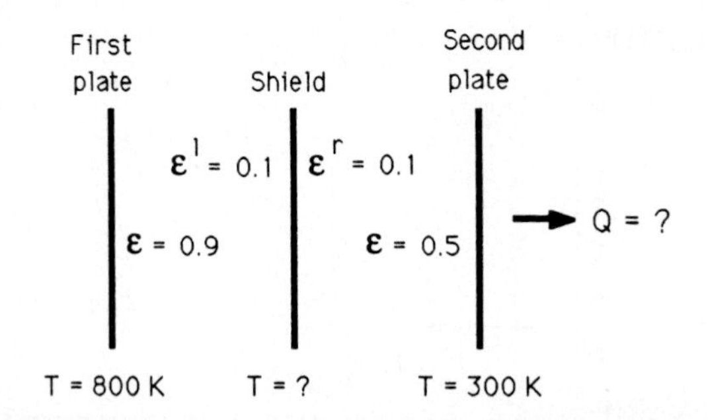

Solution. This example is similar to the previous example, except it has to be solved twice, first without the radiation shield and then with the radiation shield.

Table below shows that, for the case without the radiation shield, the number of shields is taken as zero. For such a case the computer does not accept any INPUT data for the shield. Only INPUT data are the temperature and emissivity data for the first and second plates.

Tables below show the computer INPUT and OUTPUT data for the cases without and with the radiation shield, respectively.

```
                    RADIATION SHIELDS
                    Geometry : Parallel plates
                        Area : m²                          1.000
     First plate temperature : K                         800.000
      First plate emissivity :                             0.900

          Number of shields : 0

Emissivity of     Emissivity of
First surface     Second surface                     Temperatures =




      Last plate temperature : K                         300.000
       Last plate emissivity :                             0.500

             Heat flow rate = W                        10783.445
```

```
                    RADIATION SHIELDS
                    Geometry : Parallel plates
                        Area : m²                          1.000
     First plate temperature : K                         800.000
      First plate emissivity :                             0.900

          Number of shields : 1

Emissivity of     Emissivity of
First surface     Second surface                     Temperatures =
 0.100               0.100                          K      682.757




      Last plate temperature : K                         300.000
       Last plate emissivity :                             0.500

             Heat flow rate = W                         1078.344
```

The computer OUTPUTS show that the heat flow rates per meter square are Q=10783.445W and Q=1078.34W, respectively, for the cases without and with the radiation shield. The radiation shield reduced the heat flow rate by a factor of ten. The shield temperature is 682.757K.

Example 13-3. Consider two long coaxial cylinders as illustrated in figure below. The inner cylinder has a diameter D_1=5 cm, emissivity ε_1=0.9 and temperature T_1=450K. The outer cylinder has a diameter D_2=10 cm, emissivity ε_2=0.1 and temperature T_2=300K. Calculate the heat flow rate through this system per meter length of the cylinder.

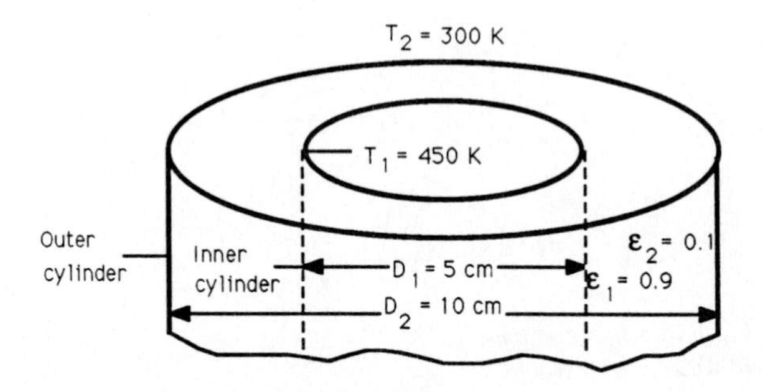

Solution. Table below shows the computer INPUT and OUTPUT data after the program has been run.

```
              RADIATION SHIELDS
              Geometry : Long cylinders
                Length : m                         1.000
 Inner cylinder temperature : K                  450.000
  Inner cylinder emissivity :                      0.900
    Inner cylinder diameter : cm                   5.000
         Number of shields : 0

Emissivity of    Emissivity of
Inner surface    Outer surface      Diameter      Temperatures =



 Outer cylinder temperature : K                  300.000
  Outer cylinder emissivity :                      0.100
    Outer cylinder diameter : cm                  10.000
            Heat flow rate = W                    52.231
```

Example 13-4. Consider a radiation shield placed between two concentric cylinders as illustrated in figure below. The *inner* cylinder has a diameter D_1=5 cm emissivity ε_1=0.9 and temperature T_1=1000K. The *outer* cylinder has a diameter D_2=10 cm emissivity ε_2=0.8 and temperature T_2=500K. The *shield* has a diameter D_s=7 cm, emissivity ε^l=0.05 for the left surface and ε^r=0.1 for the right surface. Calculate the heat transfer rate per meter length of this system without and with the radiation shield.

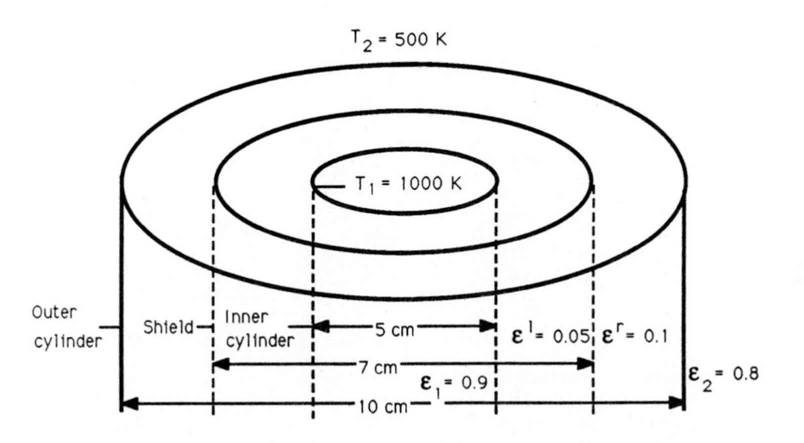

Solution. Table below show the computer INPUT-OUTPUT data after the program has been run for the cases without and with the radiation shield.

```
                     RADIATION SHIELDS
                     Geometry : Long cylinders
                       Length : m                           1.000
  Inner cylinder temperature : K                         1000.000
   Inner cylinder emissivity :                              0.900
     Inner cylinder diameter : cm                           5.000
           Number of shields : 0

Emissivity of    Emissivity of
Inner surface    Outer surface       Diameter       Temperatures =




  Outer cylinder temperature : K                          500.000
   Outer cylinder emissivity :                              0.800
     Outer cylinder diameter : cm                          10.000
              Heat flow rate = W                         6754.865
```

The INPUT data in Table above, without the radiation shield, are entered as described in the previous example. The HEAT FLOW RATE per meter length of the cylinder is Q=6754.865W.

```
                RADIATION SHIELDS
                Geometry : Long cylinders
                  Length : m                          1.000
   Inner cylinder temperature : K                  1000.000
    Inner cylinder emissivity :                       0.900
      Inner cylinder diameter : cm                    5.000
            Number of shields : 1

Emissivity of     Emissivity of
Inner surface     Outer surface        Diameter          Temperatures =
 0.050              0.100          cm         7.000  K        781.450




   Outer cylinder temperature : K                   500.000
    Outer cylinder emissivity :                       0.800
      Outer cylinder diameter : cm                   10.000
             Heat flow rate = W                     380.392
```

The *Heat flow rate* with the shield is 380.392W. Clearly, the presence of radiation shield reduced the heat flow rate to approximately one seventeenth of that without the shield.

Example 13-5. Consider two concentric spheres. The inner cylinder has a diameter D_1=5 cm, emissivity ε_1=0.9 and temperature T_1=800K. The outer cylinder has a diameter D_2=10 cm, emissivity ε_2=0.8 and temperature T_2=300K. Figure below show the geometry. Calculate the heat transfer rate through this system.

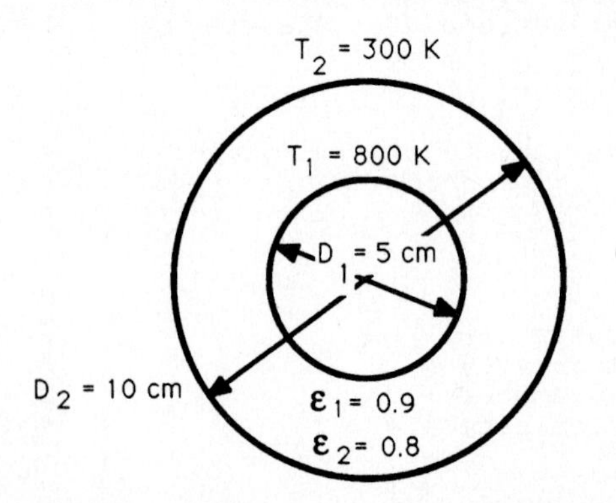

Solution. Table below shows the computer INPUT and OUTPUT data for this example after the program has been run.

```
                    RADIATION SHIELDS
                    Geometry : Concentric spheres

  Inner sphere temperature : K                      800.000
   Inner sphere emissivity :                          0.900
     Inner sphere diameter : cm                       5.000
         Number of shields : 0

Emissivity of    Emissivity of
Inner surface    Outer surface    Diameter    Temperatures =




  Outer sphere temperature : K                      300.000
   Outer sphere emissivity :                          0.800
     Outer sphere diameter : cm                      10.000
           Heat flow rate = W                       152.347
```

There is no shield for this problem, so the *Number of shields* is set to zero. As a result the computer does not accept any emissivity data for the shield.

Example 13-6. Consider Example 13-5 with two shields placed between the spheres as shown in the figure below. One of the shields has a diameter 7 cm,

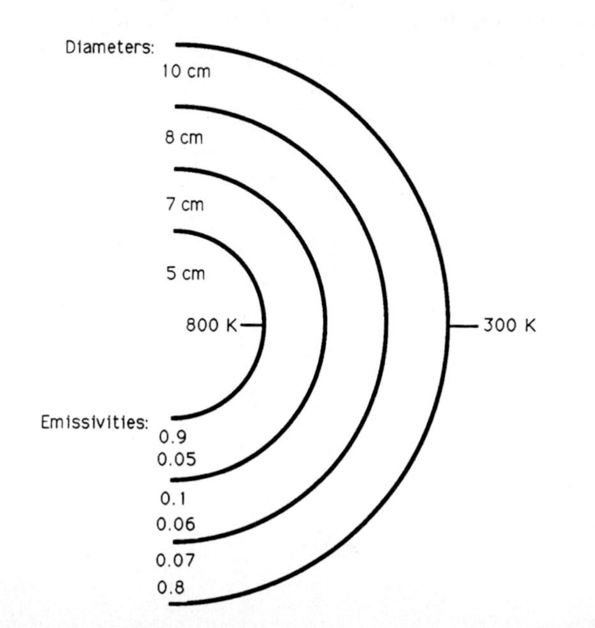

emissivities 0.05 for its inner surface and 0.1 for its outer surface. The other shield has a diameter 8 cm, emissivities 0.06 for the inner surface and 0.07 for the outer surface. Figure above shows the geometry. Calculate the heat flow rate.

Solution. Table below shows the computer INPUT and OUTPUT data for this example after the program has been run.

```
                    RADIATION SHIELDS
                    Geometry : Concentric spheres

     Inner sphere temperature : K                    800.000
      Inner sphere emissivity :                        0.900
        Inner sphere diameter : cm                     5.000
            Number of shields : 2

Emissivity of    Emissivity of
Inner surface    Outer surface      Diameter        Temperatures =
 0.050             0.100          cm       7.000  K       709.025
 0.060             0.0702         cm       8.000  K       547.413


     Outer sphere temperature : K                    300.000
      Outer sphere emissivity :                        0.800
        Outer sphere diameter : cm                    10.000
              Heat flow rate = W                       6.466
```

Note that, a comparison of this heat flow rate with that in Example 13-5, shows that the addition of two shields reduced the heat flow rate from 152.347W to 6.462W.

Chapter

FOURTEEN

HEAT EXCHANGER EFFECTIVENESS

Heat exchangers are devices that facilitate heat transfer between two or more fluids at different temperatures. Many types of heat exchangers have been developed for use at such varied levels of technological sophistication and sizes as steam power plants, chemical processing plants, building heating and air conditioning, household refrigerators, car radiators, radiators for space vehicles, and so on. In the common types, such as shell-and-tube heat exchangers and car radiators, heat transfer is primarily by conduction and convection from a hot to a cold fluid, which are separated by a metal wall. In boilers and condensers,heat transfer by boiling and condensation is of primary importance. In radiators for space applications, the waste heat carried by the coolant fluid is transported by convection and conduction to the fin surface and from there by thermal radiation into the atmosphere-free space.

The design of heat exchangers is a complicated matter. Heat transfer and pressure drop analysis, sizing and performance estimation, and the economic aspects play important roles in the final design. For example, although the cost considerations are very important for applications in large installations such as power plants and chemical processing plants, the weight and size considerations become the dominant factor in the choice of design for space and aeronautical applications. A comprehensive treatment of heat exchanger design is, therefore, beyond the scope of this heat transfer solver.

In the thermal analysis of heat exchangers, the *rating* and *sizing* are two important problems. The rating is concerned with the determination of heat transfer rate through the exchanger while the sizing is concerned with the determination of the heat transfer surface to meet a specified heat transfer rate.

In many situations, the use of the Logarithmic Mean Temperature Difference (LMTD) method of thermal analysis of heat exchangers may require tedious iterations.

The *effectiveness* method of analysis originally developed by Kays and London (1964) is very convenient to use in the solution of rating and sizing problems. Given the effectiveness, the heat transfer rate through the heat exchanger is readily calculated.

Therefore, this chapter is devoted to the determination of the effectiveness for various types of heat exchangers.

The definition of the effectiveness is given, analytic expressions for the determination of effectiveness for various types of heat exchangers are presented, the use of the program Heat Exchanger Effectiveness is illustrated with examples.

14-1 ANALYTIC EXPRESSIONS FOR EFFECTIVENESS

The effectiveness ε of a heat exchanger is defined as

$$\varepsilon = \frac{Q}{Q_{max}} = \frac{\text{actual heat transfer rate}}{\text{maximum possible heat transfer rate from one stream to the other}} \tag{14-1}$$

The maximum possible heat transfer rate Q_{max} is obtained with a counterflow exchanger if the temperature change of the fluid having the minimum value of mc_p equals the difference in the inlet temperatures of the hot and cold fluids. Here we consider $(mc_p)_{min}$, because the energy given up by one fluid should equal that received by the other fluid. If we consider $(mc_p)_{max}$, then the other fluid should undergo a temperature change greater than the maximum available temperature difference; that is, ΔT for the other fluid should be greater than $T_{h,in} - T_{c,in}$. This is not possible. With this consideration, Q_{max} is chosen as

$$Q_{max} = (mc_p)_{min}(T_{h,in} - T_{c,in}) \tag{14-2}$$

Then, given ε and Q_{max}, the actual heat transfer rate Q is

$$Q = \varepsilon(mc_p)_{min}(T_{h,in} - T_{c,in}) \tag{14-3}$$

Here $(mc_p)_{min}$ is the smaller of m_hc_{ph} and m_cc_{pc} for the hot and cold fluids; $T_{h,in}$ and $T_{c,in}$ are the inlet temperatures of the hot and cold fluids, respectively.

Clearly, if the effectiveness ε of the heat exchanger is known, Eq. (14-3) provides an explicit expression for the determination of heat transfer rate Q through the exchanger. The reader may refer to the text by Ozisik (1985) for detailed discussion on the use of effectiveness method of analysis of rating and sizing of heat exchangers.

For a given type of heat exchanger, the effectiveness ε is a function of two quantities N and C defined as

$$N = \frac{UA}{C_{min}} \equiv \text{Number of heat transfer units} \tag{14-4a}$$

$$C = \frac{C_{min}}{C_{max}} \equiv \text{Heat capacity ratio} \tag{14-4b}$$

where

U = overall heat transfer coefficient

A = heat transfer surface area

C_{min} and C_{max} = the smaller and the larger, respectively, of the two quantities C_h and C_c

C_h = $m_h c_{ph}$

C_c = $m_c c_{pc}$

c_{pc}, c_{ph} = specific heat of the cold and hot fluid, respectively

m_c, m_h = mass flow rate (i.e., kg/s) of the cold and hot fluid, respectively

Therefore, the effectiveness ε of a given type of heat exchanger can be expressed formally, as a function of N and C, in the form

$$\varepsilon = f(N,C) \tag{14-5a}$$

which provides a convenient relationship between ε and N. For example, given the heat transfer surface area A, U, C and C_{min}, the effectiveness is readily calculated.

Suppose we rearrange Eq. (14-5a), if possible, in the form

$$N = f^*(\varepsilon, C) \tag{14-5b}$$

Then, for a given ε and C, the parameter N is known. Then, given U and C_{min}, the heat transfer area A is determined from the definition

$$N = \frac{UA}{C_{min}} \tag{14-5c}$$

The above consideration leads to the analysis of the following two different types of problems:

(1) The effectiveness ε is given, the heat transfer area A of the heat exchanger is to be determined.

(2) The heat transfer area A is given, the effectiveness ε of the heat exchanger is to be determined.

In the following analysis, we consider three different types of heat exchangers, and for each of them present analytic expressions in the forms

$$\varepsilon = f(N,C) \tag{14-6a}$$

and

$$N = f^*(\varepsilon,C) \tag{14-6b}$$

Double Pipe Heat Exchangers

Figure (14-1) show schematically the flow arrangements for a *parallel flow* and a *counter flow* double pipe heat exchanger, respectively.

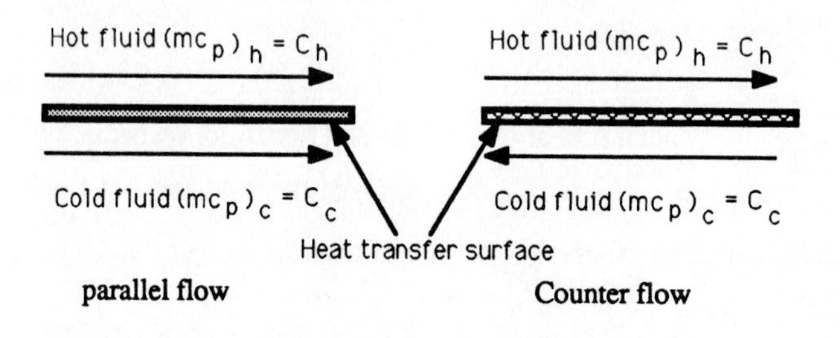

Figure 14-1 Nomenclature for double pipe heat exchangers

The expressions defining $\varepsilon = f(N,C)$ and $N = f^*(\varepsilon,C)$ are given by

(a) Parallel flow.

$$\varepsilon = \frac{1-e^{-N(1+C)}}{1 + C} \tag{14-7a}$$

and

$$N = -\frac{\ln[1 - \varepsilon(1+C)]}{1 + C} \qquad \text{for } \varepsilon \leq \frac{1}{1+C} \tag{14-7b}$$

(b) Counter flow.

$$\varepsilon = \begin{cases} \dfrac{1-e^{-N(1-C)}}{1-C\,e^{-N(1-C)}} & \text{for } C < 1 \quad (14\text{-}8a) \\[2ex] \dfrac{N}{1+N} & \text{for } C = 1 \quad (14\text{-}8b) \end{cases}$$

and

$$N = \begin{cases} \dfrac{1}{C-1}\ln\left(\dfrac{1-\varepsilon}{1-\varepsilon C}\right) & \text{for } \varepsilon < 1 \quad (14\text{-}9a) \\[2ex] \dfrac{\varepsilon}{1 - \varepsilon} & \text{for } C = 1 \quad (14\text{-}9b) \end{cases}$$

Shell-and-Tube Heat Exchangers

We consider the following two types of shell-and-tube heat exchangers:

(i) Single shell pass with 2, 4, 6, ... etc. tube passes,

(ii) n-shell pass with 2n, 4n, 6n, ... etc. tube passes.

Figure 14-2 illustrates a single shell pass two tube pass heat exchanger.

The expressions defining $\varepsilon = f(N,C)$ and $N = f^{*}(\varepsilon,C)$ are given by

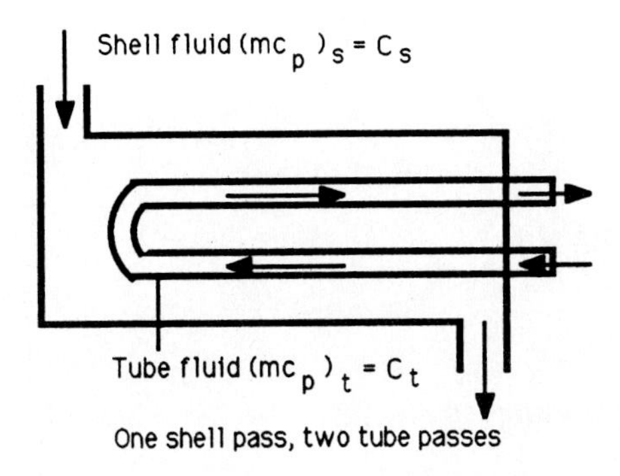

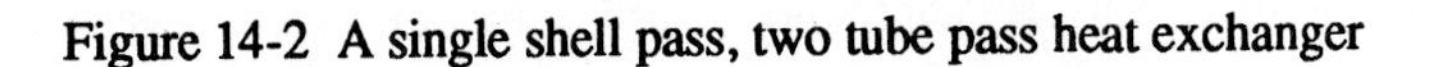
Figure 14-2 A single shell pass, two tube pass heat exchanger

(a) Single shell pass with 2, 4, 6, etc. tube passes:

$$\varepsilon = 2\left\{1 + C + (1+C^2)^{1/2}\,\frac{1 + e^{-N(1+C^2)^{1/2}}}{1 - e^{-N(1+C^2)^{1/2}}}\right\}^{-1} \tag{14-10}$$

and

$$N = -\frac{1}{(1 + C^2)^{1/2}}\ln\left(\frac{E-1}{E+1}\right) \tag{14-11a}$$

where

$$E = \frac{\frac{2}{\varepsilon} - (1+C)}{(1 + C^2)^{1/2}} \tag{14-11b}$$

(b) n-shell pass with 2n, 4n, 6n, etc. tube passes

$$\varepsilon = \left[\left(\frac{1 - \varepsilon_1 C}{1 - \varepsilon_1}\right)^n - 1\right]\left[\left(\frac{1 - \varepsilon_1 C}{1 - \varepsilon_1}\right)^n - C\right]^{-1} \quad \text{for } C < 1 \tag{14-12a}$$

$$\varepsilon = \frac{n}{n - 1 + (1/\varepsilon_1)} \quad \text{for } C = 1 \tag{14-12b}$$

and

$$N = -\frac{1}{(1 + C^2)^{1/2}} \ln\left(\frac{E - 1}{E + 1}\right) \tag{14-12c}$$

where

$$E = \frac{\frac{2}{\varepsilon_1} - (1 + C)}{(1 + C)^{1/2}} \tag{14-12d}$$

$$\varepsilon_1 = \frac{\varepsilon}{n + \varepsilon - n\varepsilon} \quad \text{for} \quad C = 1 \tag{14-12e}$$

All heat exchangers with C=0:

$$\varepsilon = 1 - e^{-N} \tag{14-13}$$

Cross-Flow Heat Exchangers

We consider cross-flow heat exchangers for the following flow arrangements.

	C_{max}	C_{min}
(a)	mixed	unmixed
(b)	unmixed	mixed
(c)	unmixed	unmixed
(d)	mixed	mixed

The reader should refer to Ozisik (1985) p. 538 for a discussion of the concept of mixed and unmixed flow arrangements.

Figure 14-3 illustrates a cross-flow arrangement.

We present below the expressions for ε and N for each of these four different types of flow arrangements.

(a) C_{max}(mixed), C_{min} (mixed).

$$\varepsilon = \frac{1}{C}\{1 - \exp[-C(1 - e^{-N})]\} \tag{14-14}$$

and

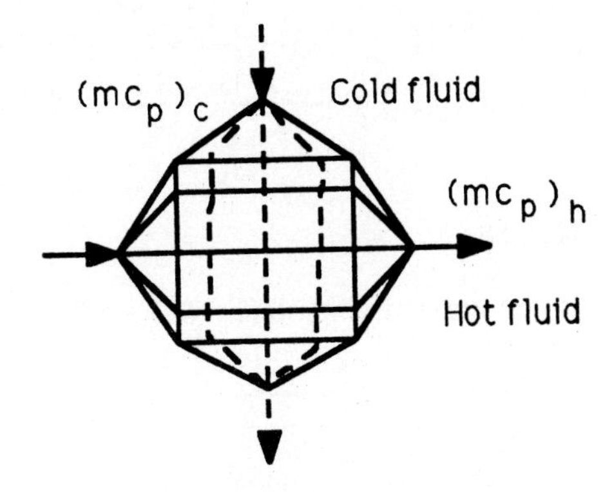

Figure 14-3 A flow arrangement in a cross-flow heat exchanger

$$N = -\ln\left\{1 + \frac{1}{C}\ln(1 - \varepsilon C)\right\} \quad \text{for} \quad \varepsilon < \frac{1}{C}(1 - e^{-C}) \qquad (14\text{-}15)$$

(b) C_{max}(unmixed), C_{min}(mixed).

$$\varepsilon = 1 - \exp\left\{-\frac{1}{C}(1 - e^{-NC})\right\} \qquad (14\text{-}16)$$

and

$$N = -\frac{1}{C}\ln\{1 + C\ln(1 - \varepsilon)\} \quad \text{for} \quad \varepsilon < 1 - e^{-1/c} \qquad (14\text{-}17)$$

(c) C_{max}(unmixed), C_{min}(mixed).

$$\varepsilon = 1 - \exp\left[\frac{N^{0.22}}{C}\{\exp(-CN^{0.78}) - 1\}\right] \qquad (14\text{-}18)$$

and

$$N_{k+1} = \left\{-\frac{1}{C}\ln\left[1 + \frac{C}{N_k^{0.22}}\ln(1 - \varepsilon)\right]\right\}^{1/0.78} \qquad (14\text{-}19)$$

where $|N_{k+1} - N_k| < \varepsilon$

and the initial guess $N_o = -\ln(1 - \varepsilon)$.

(d) C_{max}(mixed), C_{min}(mixed)

$$\varepsilon = \left[\frac{1}{1 - e^{-N}} + \frac{C}{1 - e^{-NC}} - \frac{1}{N}\right]^{-1} \qquad (14\text{-}20)$$

and

$$N_{k+1} = -\ln\left[1 - \frac{1 + C\,\dfrac{1 - e^{-N_k}}{1 - e^{-N_k C}}}{\dfrac{1}{N_k} + \dfrac{1}{\varepsilon}}\right] \tag{14-21}$$

where $|N_{k+1} - N_k| < \varepsilon$

initial guess $N_o = -\ln[1 - \varepsilon(1 - C)]$

$$\varepsilon \leq \frac{1}{1 + C}$$

The foregoing expressions are the basis for calculation of heat exchanger effectiveness ε or the quantity N.

14-2 COMPUTER SOLUTIONS

The foregoing expressions for ε and N are used in the program *HEAT EXCHANGER EFFECTIVENESS* in order to solve the following types of problems. Assume that the following INPUT data are available:

Hot fluid flow rate $\equiv$ m_h (kg/s)
Cold fluid flow rate $\equiv$ m_c (kg/s)
Hot fluid specific heat $\equiv$ c_{ph} (J/kg·K)
Cold fluid specific heat $\equiv$ c_{pc} (J/kg·K)
Overall heat transfer coefficient $\equiv$ U (W/m^2·K)

The program *HEAT EXCHANGER EFFECTIVENESS* can be used to determine:

(1) The effectiveness ε, when the heat transfer area A is given,

(2) The heat transfer area A, when the effectiveness ε is specified.

We illustrate the application with representative examples.

Example 14-1. A *counterflow double pipe* heat exchanger of heat transfer area A=12.5 m^2 is to cool oil [c_{ph} = 2000 J/(kg · s)] with water [c_{pc} = 4170 J/(kg · s)]. The oil enters at $T_{h,in}$ = 100°C and m_h = 2 kg/s, while the water enters at $T_{c,in}$ = 20°C and m_c = 0.48 kg/s. The overall heat transfer coefficient is U_m = 400 W/(m^2· °C). Calculate the heat exchanger effectiveness.

Solution. Table below shows the computer INPUT and OUTPUT data for this example after the program has been run.

```
                    HEAT EXCHANGER EFFECTIVENESS

                 Exchanger type :  Double pipe
              Flow arrangement :  Counter flow

           Hot fluid flow rate :  kg/s                 2.000
          Cold fluid flow rate :  kg/s                 0.480
       Hot fluid specific heat :  J/kg•K            2000.000
      Cold fluid specific heat :  J/kg•K            4170.000
Overallheat transfer coefficient :  W/m²•K             400.000
                  Surface area :  m²                  12.500


                 Effectiveness =                       0.833
```

We now describe the entering the INPUT data.

The first item is the selection of the *Exchanger type.* Move the cursor next to the Exchanger type and press the SPACEBAR. Each time SPACEBAR is pressed, the type changes, successively, to *Double pipe, Shell and tube* and *Crossflow.* Press the ENTER key when *Double pipe* is displayed.

The next item is the *Flow arrangement.* Move the cursor next to the flow arrangement and press the SPACEBAR. Each time it is pressed, the flow arrangement alternate between *Parallel flow* and *Counter flow.* Press the ENTER key when *Counter flow* is displayed.

The next five items are *Hot fluid flow rate, Cold fluid flow rate, Hot fluid specific heat, Cold fluid specific heat* and the *Overall heat transfer coefficient.* Enter first the unit and then the values for each of them.

The next two items are for the *Surface area* and *Effectiveness.* They can be interchanged and the last one with the equal sign "=" becomes the computer OUTPUT (i.e., unknown). Move the cursor over the *Surface area* and press the SPACEBAR. Each time it is pressed, it alternates between *Surface area* and *Effectiveness.* Press the ENTER key when the *Surface area* is displayed, because the area is given in this problem. Then the *Effectiveness* appear in the last item as the OUTPUT.

Run the program by pressing the F5 function key; the solution gives

Effectiveness = 0.833.

Graphical Support. Using the graphical support, the effectiveness ε can be plotted against the heat transfer surface area A as shown in the figure below.

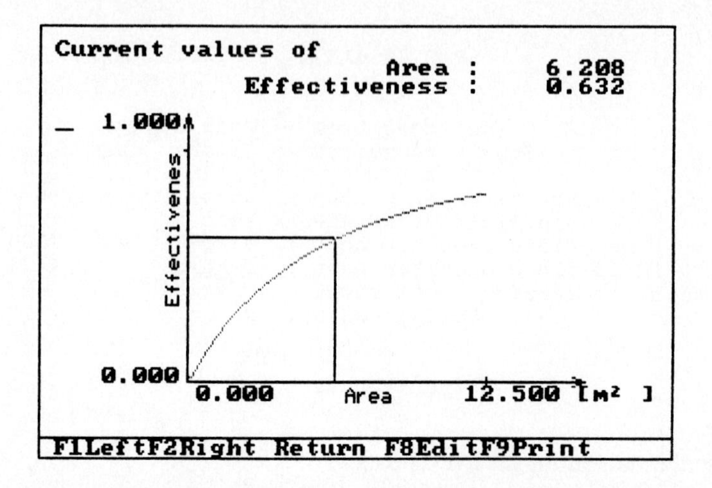

Each time the ENTER key is pressed, the current values of X and Y are stored in the computer memory up to 150 such results can be stored in the memory, since the calculations are performed at 150 equally divided locations over the selected interval from X=5 to X=20 m^2. Results stored in the memory can be printed on the printer by pressing F9 function key.

Example 14-2. Repeat Example 14-1 by assuming everything exactly the same, except the effectiveness is given as ε=0.833 and the surface area A is to be determined.

Solution. Table Ex14-2 shows the computer INPUT and OUTPUT data for this example after the program has been run.

HEAT EXCHANGER EFFECTIVENESS

Exchanger type	:	Double pipe	
Flow arrangement	:	Counter flow	
Hot fluid flow rate	:	kg/s	2.000
Cold fluid flow rate	:	kg/s	0.480
Hot fluid specific heat	:	J/kg•K	2000.000
Cold fluid specific heat	:	J/kg•K	4170.000
Overall heat transfer coefficient	:	W/m²•K	400.000
Effectiveness	:		0.833
Surface area	=	m²	12.525

A comparison of the results shown in the above Tables reveals that, the first seven of the INPUT data are exactly the same, but the last two items have their locations interchanged.

That is, in Table of Example 14-1 the *Effectiveness* followed by the equal sign is the last item that represents the OUTPUT, whereas in Table of Example 14-2 the *Surface* area followed by the equal sign is the last item that represents the OUTPUT.

Move the cursor over the item before the last one. It may be the *Effectiveness* or the *Surface area.* Each time the SPACEBAR is pressed, the *Effectiveness* and *Surface area* alternates. When the desired quantity appears (i.e., effectiveness for this example) press the ENTER key. This locks the *Effectiveness* and simultaneously causes the last item to change to *Surface area.* Enter the value of the effectiveness and run the program by pressing F5 function key. The computer output gives

surface area = 12.525

The slight difference between the values of surface area of 12.525 in this example and 12.5 in the previous one can be eliminated if one more significant figure is retained in the value of effectiveness, i.e., taking $\varepsilon=0.8325$ we obtain $A=12.499\ m^2$.

Example 14-3. A one-shell and two-tube pass heat exchanger will heat cold water flowing at a rate of $m_c=2$ kg/s with pressurized hot water flowing at a rate of $m_h=2$ kg/s. The overall heat transfer coefficient is $U=1250\ W/m^2{\cdot}K$, the heat exchanger effectiveness is $\varepsilon=0.385$ and the heat capacity rates are $C_{ph}=C_{pc}=4180$ J/kg·K. Calculate the heat transfer surface A.

Solution. Table below shows the computer INPUT and OUTPUT data for this example after the program has been run.

HEAT EXCHANGER EFFECTIVENESS

Exchanger type	:	Shell and tube	
Number of shell passes	:	1	
Number of tube passes	:	2,4,6,8 ...	
Hot fluid flow rate	:	kg/s	2.000
Cold fluid flow rate	:	kg/s	2.000
Hot fluid specific heat	:	J/kg•K	4180.000
Cold fluid specific heat	:	J/kg•K	4180.000
Overall heat transfer coefficient	:	W/m²•K	1250.000
Effectiveness	:		0.385
Surface area	=	m²	4.498

The first item is the Exchanger type. Move the cursor next to the Exchanger type, press the SPACEBAR and select *Shell and tube* type as discussed previously in the Example 14-1.

Next enter 1 for the Number of Shell Passes; immediately afterwards the numbers 2,4,6,8,... appear next to the number of tube passes.

The next five items are *Hot fluid flow rate, Cold fluid flow rate, Hot fluid specific heat, Cold fluid specific heat* and the *Overall heat transfer coefficient.* These data are entered in the usual manner.

The last two items are the *Effectiveness* and the *Surface area.* They can be interchanged. The one that appears last with the equal sign is the computer OUTPUT or the solution for the problem. In this example the effectiveness is specified. Therefore enter the value of the *Effectiveness* as ε=0.385.

Press the F5 function key to run the program. The OUTPUT gives

Surface area = 4.498 m^2.

Example 14-4. A cross-flow heat exchanger with both fluids unmixed will heat water with hot exhaust gas. The flow rates of water and the exhaust gas are, respectively, m_c=1.5 kg/s and m_h=2.5 kg/s. The specific heats of water and exhaust gas are respectively, c_{pc}=4180 J/kg·K and c_{ph}=1050 J/kg·K. The overall heat transfer coefficient is U=150 W/m^2·K and the exchanger effectiveness is ε=0.558. Calculate the heat transfer surface A.

Solution. Table below shows the computer INPUT and OUTPUT data for this example after the program has been run.

HEAT EXCHANGER EFFECTIVENESS

Exchanger type	:	Cross flow	
Hot fluid	:	Unmixed	
Cold fluid	:	Unmixed	
Hot fluid flow rate	:	kg/s	2.500
Cold fluid flow rate	:	kg/s	1.500
Hot fluid specific heat	:	J/kg•K	1050.000
Cold fluid specific heat	:	J/kg•K	4180.000
Overall heat transfer coefficient	:	W/m²•K	150.000
Effectiveness	:		0.558
Surface area	=	m²	17.485

The first item is the Exchanger Type which is selected as *Cross flow* as explained previously in Example 14-1.

Move the cursor next to the *Hot fluid* and press the SPACEBAR. Each time the SPACEBAR is pressed, screen alternately displays *Mixed* and *Unmixed.* Press the ENTER key when *Unmixed* appears, since in this problem the hot fluid is *Unmixed.*

Similarly, choose *Unmixed* for the cold fluid.

The next five items are for the *Flow rate* and *Specific heat* of *Hot* and *Cold* fluids and the *Overall heat transfer coefficient*. Enter these data as discussed previously.

The last two items, the *Effectiveness* and *Surface area* can be interchanged. The one that appears last with the equal sign is the OUTPUT and answer to the problem.

In this Example, the *Effectiveness* is given by ε=0.558, which is entered as discussed previously.

The computer is run by pressing the F9 function key and computer OUTPUT gives the solution as

$$\textit{Surface area} = 17.485 \text{ m}^2.$$

Example 14-5. Repeat Example 14-4 by assuming everything remains exactly the same, except the surface area A=17.485 m^2 is given and the effectiveness ε is to be determined.

Solution. Table below shows the computer INPUT and OUTPUT data for this example after the program has been run.

HEAT EXCHANGER EFFECTIVENESS

Exchanger type	:	Cross flow	
Hot fluid	:	Unmixed	
Cold fluid	:	Unmixed	
Hot fluid flow rate	:	kg/s	2.500
Cold fluid flow rate	:	kg/s	1.500
Hot fluid specific heat	:	J/kg•K	1050.000
Cold fluid specific heat	:	J/kg•K	4180.000
Overall heat transfer coefficient	:	W/m²•K	150.000
Surface area	:	m²	17.485

Effectiveness = 0.558

The effectiveness is determined as ε=0.558. A comparison of this result with that of the previous example shows that the effectiveness for both cases are the same as expected.

REFERENCES

Amato, W. S. and C. Tien, Free Convection Heat Transfer from Isothermal Sphere in Water, Int. J. Heat Mass Transfer, **15**, 327-339, 1972.

Bromley, L. A., Heat Transfer in Stable Film Boiling, Chem. Eng. Progress, **46**, 221-227, 1950.

Carslaw, H. S. and J. C. Jaeger, Conduction of Heat in Solids, 2nd ed., Oxford at the Clarendon Press London, 1959.

Churchill, S. W. and H. H. S. Chu, Correlating Equations for Laminar and Turbulent Free Convection from a Vertical Plate, Int. J. Heat Mass Transfer, **18**, 1323-1329, 1975.

Churchill, S. W. and H. H. S. Chu, Correlating Equations for Laminar and Free Convection from a Horizontal Cylinder, Int. J. Heat Mass Transfer, **18**, 1049-1053, 1975.

Churchill, S. W. and M. Bernstein, A. Correlating Equation for Forced Convection from Gases and Liquids to a Circular Cylinder in Cross Flow, J. Heat Transfer, **99**, 300-306, 1977.

Ded, J. S. and J. H. Lienhard, The Peak Boiling Heat Flux from a Sphere, A.I.Ch.E., **18**, 337-342, 1972.

Farber, E. and R. L. Scorah, Heat Transfer to Water Boiling Under Pressure, Trans. ASME, **79**, 369-384, 1948.

Gardner, K. A., Efficiency of Extended Surfaces, Trans. ASME, **67**, 621-631, 1945.

Hamilton, D. C. and W. R. Morgan, Radiation Interchange Configuration Factors, NACA Tech. Note. 2836, 1952.

Harper, W. P. and D. R. Brown, Mathematical Equations for Heat Conduction in the fins of Air-Cooled Engines, NACA Rep. 158, 1922.

Howell, J. R., A Catalog of Radiative Angle Factors, McGraw-Hill, New York, 1982.

Kays, W. M. and A. L. London, Compact Heat Exchangers, 2nd ed., McGraw-Hill, New York, 1964.

Kern, D. Q. and A. D. Kraus, Extended Surface Heat Transfer, McGraw-Hill, New York, 1972.

Kirkbride, C. G., Heat Transfer by Condensing Vapor, Trans. A.I.Che.E., **30**, 170-186, 1934.

Lienhard, J. H. and V. K. Dhir, Hydrodynamic Prediction of Peak Pool-Boiling Heat Fluxes from Finite Bodies, J. Heat Transfer, **95C**, 152-158, 1973.

Mackey, C. O., L. T. Wright, R. E. Clark, and N. R. Gray, Radiant Heating and Cooling. Pt. I., Cornell Univ. Eng. Exp. Sta. Bull., **22**, 1943.

Mikhailov, M. D. and M. N. Özisik, Unified Analysis and Solution of Heat and Mass Diffusion, Wiley, New York, 1984.

Notter, R. H. and C. A. Sleicher, A Solution to Turbulent Graetz Problem. III. Fully Developed and Entry Heat Transfer Rates, Chem. Eng. Sci., **27**, 2073-2093, 1972.

Mikhailov, M. D. and B. K. Shishedjiev, Chebishev Approximation for the Roots of the Equation BiWG(m) = mVG(m), Journal of Engineering Physics, **28**, 509-513, 1975.

Nusselt, W., Die Ober flachenkondensation des Wasserdampfes, Z. Ver. Deut. Ing., **60**, 541-569, 1916.

Özişik, M. N., Heat Conduction, Wiley, New York, 1980.

Özişik, M. N., Heat Transfer, McGraw-Hill, New York, 1985.

Planck, M., Theory of Heat Radiation, Dover, New York, 1959.

Rohsenow, W. M., A. Method of Correlating Heat Transfer Data for Surface Boiling Liquids, Trans. ASME, **74**, 969-975, 1952.

Siegel, R. and J. R. Howell, Thermal Radiation Heat Transfer, Hemisphere, New York, 1981.

Sun, K. H. and J. H. Lienhard, The Peak Boiling Heat Flux on Horizontal Cylinders, Int. J. Heat Mass Transfer, **13**, 1425-1439, 1970.

Whitaker, S., Forced Convection Heat Transfer Calculations for Flow in Pipes, Past Flat Plates, Single Cylinders, and Flow in Packed Beds and Tube Bundles, AIChE J, **18**, 361-371, 1972.

Zukauskas, A., Heat Transfer from Tubes in Cross-Flow, Adv. Heat Transfer, **8**, 93-160, 1972.

Additional Notes:

Additional Notes:

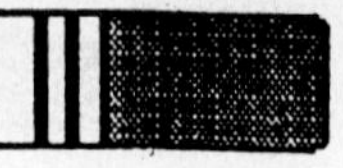

Additional Notes:

Additional Notes:

Additional Notes:

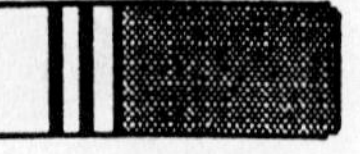

Additional Notes:

HEAT TRANSFER SOLVER
M.D. Mikhailov • M.N. Özisik

YOU SHOULD CAREFULLY READ THE FOLLOWING TERMS AND CONDITIONS BEFORE OPENING THIS DISKETTE PACKAGE. OPENING THIS DISKETTE PACKAGE INDICATES YOUR ACCEPTANCE OF THESE TERMS AND CONDITIONS. IF YOU DO NOT AGREE WITH THEM, YOU SHOULD PROMPTLY RETURN THE PACKAGE UNOPENED.

Prentice-Hall, Inc. provides this program and licenses its use. You assume responsibility for the selection of the program to achieve your intended results, and for the installation, use, and results obtained from the program. This license extends only to use of the program in the United States or countries in which the program is marketed by duly authorized distributors.

LICENSE

You may:

a. use the program;
b. copy the program into any machine-readable form without limit;
c. modify the program and/or merge it into another program in support of your use of the program.

LIMITED WARRANTY

THE PROGRAM IS PROVIDED "AS IS" WITHOUT WARRANTY OF ANY KIND, EITHER EXPRESSED OR IMPLIED, INCLUDING, BUT NOT LIMITED TO, THE IMPLIED WARRANTIES OF MERCHANTABILITY AND FITNESS FOR A PARTICULAR PURPOSE. THE ENTIRE RISK AS TO THE QUALITY AND PERFORMANCE OF THE PROGRAM IS WITH YOU. SHOULD THE PROGRAM PROVE DEFECTIVE, YOU (AND NOT PRENTICE-HALL, INC. OR ANY AUTHORIZED DISTRIBUTOR) ASSUME THE ENTIRE COST OF ALL NECESSARY SERVICING, REPAIR, OR CORRECTION.

SOME STATES DO NOT ALLOW THE EXCLUSION OF IMPLIED WARRANTIES, SO THE ABOVE EXCLUSION MAY NOT APPLY TO YOU. THIS WARRANTY GIVES YOU SPECIFIC LEGAL RIGHTS AND YOU MAY ALSO HAVE OTHER RIGHTS THAT VARY FROM STATE TO STATE.

Prentice-Hall, Inc. does not warrant that the functions contained in the program will meet your requirements or that the operation of the program will be uninterrupted or error free.

However, Prentice-Hall, Inc., warrants the diskette(s) on which the program is furnished to be free from defects in materials and workmanship under normal use for a period of ninety (90) days from the date of delivery to you as evidenced by a copy of your receipt.

LIMITATIONS OF REMEDIES

Prentice-Hall's entire liability and your exclusive remedy shall be:

1. the replacement of any diskette not meeting Prentice-Hall's "Limited Warranty" and that is returned to Prentice-Hall, or
2. if Prentice-Hall is unable to deliver a replacement diskette or cassette that is free of defects in materials or workmanship, you may terminate this Agreement by returning the program.

IN NO EVENT WILL PRENTICE-HALL BE LIABLE TO YOU FOR ANY DAMAGES, INCLUDING ANY LOST PROFITS, LOST SAVINGS, OR OTHER INCIDENTAL OR CONSEQUENTIAL DAMAGES ARISING OUT OF THE USE OR INABILITY TO USE SUCH PROGRAM EVEN IF PRENTICE-HALL OR AN AUTHORIZED DISTRIBUTOR HAS BEEN ADVISED OF THE POSSIBILITY OF SUCH DAMAGES, OR FOR ANY CLAIM BY ANY OTHER PARTY.

SOME STATES DO NOT ALLOW THE LIMITATION OR EXCLUSION OF LIABILITY FOR INCIDENTAL OR CONSEQUENTIAL DAMAGES, SO THE ABOVE LIMITATION OR EXCLUSION MAY NOT APPLY TO YOU.

GENERAL

You may not sublicense, assign, or transfer the license or the program except as expressly provided in this Agreement. Any attempt otherwise to sublicense, assign, or transfer any of the rights, duties, or obligations hereunder is void.

This Agreement will be governed by the laws of the State of New York.

Should you have any questions concerning this Agreement, you may contact Prentice-Hall, Inc., by writing to:

Prentice Hall
College Division
Englewood Cliffs, N.J. 07632

Should you have any questions concerning technical support you may write to:

M.N. Özisik
Department of Mechanical &
Aerospace Engineering
North Carolina State University
Box 7910
Raleigh, NC 27695-7910

YOU ACKNOWLEDGE THAT YOU HAVE READ THIS AGREEMENT, UNDERSTAND IT, AND AGREE TO BE BOUND BY ITS TERMS AND CONDITIONS. YOU FURTHER AGREE THAT IT IS THE COMPLETE AND EXCLUSIVE STATEMENT OF THE AGREEMENT BETWEEN US THAT SUPERCEDES ANY PROPOSAL OR PRIOR AGREEMENT, ORAL OR WRITTEN, AND ANY OTHER COMMUNICATIONS BETWEEN US RELATING TO THE SUBJECT MATTER OF THIS AGREEMENT.

ISBN 0-13-388802-9